Dawn E. Holmes, Lakhmi C. Jain (Eds.)

Innovations in Machine Learning

Studies in Fuzziness and Soft Computing, Volume 194

Editor-in-chief
Prof. Janusz Kacprzyk
Systems Research Institute
Polish Academy of Sciences
ul. Newelska 6
01-447 Warsaw
Poland
E-mail: kacprzyk@ibspan.waw.pl

Further volumes of this series
can be found on our homepage:
springer.com

Vol. 179. Mircea Negoita,
Bernd Reusch (Eds.)
*Real World Applications of Computational
Intelligence*, 2005
ISBN 3-540-25006-9

Vol. 180. Wesley Chu,
Tsau Young Lin (Eds.)
Foundations and Advances in Data Mining,
2005
ISBN 3-540-25057-3

Vol. 181. Nadia Nedjah,
Luiza de Macedo Mourelle
Fuzzy Systems Engineering, 2005
ISBN 3-540-25322-X

Vol. 182. John N. Mordeson,
Kiran R. Bhutani, Azriel Rosenfeld
Fuzzy Group Theory, 2005
ISBN 3-540-25072-7

Vol. 183. Larry Bull, Tim Kovacs (Eds.)
Foundations of Learning Classifier Systems,
2005
ISBN 3-540-25073-5

Vol. 184. Barry G. Silverman, Ashlesha Jain,
Ajita Ichalkaranje, Lakhmi C. Jain (Eds.)
*Intelligent Paradigms for Healthcare
Enterprises*, 2005
ISBN 3-540-22903-5

Vol. 185. Spiros Sirmakessis (Ed.)
Knowledge Mining, 2005
ISBN 3-540-25070-0

Vol. 186. Radim Bělohlávek, Vilém
Vychodil
Fuzzy Equational Logic, 2005
ISBN 3-540-26254-7

Vol. 187. Zhong Li, Wolfgang A. Halang,
Guanrong Chen (Eds.)
*Integration of Fuzzy Logic and Chaos
Theory*, 2006
ISBN 3-540-26899-5

Vol. 188. James J. Buckley, Leonard J.
Jowers
Simulating Continuous Fuzzy Systems, 2006
ISBN 3-540-28455-9

Vol. 189. Hans Bandemer
Mathematics of Uncertainty, 2006
ISBN 3-540-28457-5

Vol. 190. Ying-ping Chen
*Extending the Scalability of Linkage
Learning Genetic Algorithms*, 2006
ISBN 3-540-28459-1

Vol. 191. Martin V. Butz
*Rule-Based Evolutionary Online Learning
Systems*, 2006
ISBN 3-540-25379-3

Vol. 192. Jose A. Lozano, Pedro Larrañaga,
Iñaki Inza, Endika Bengoetxea (Eds.)
Towards a New Evolutionary Computation,
2006
ISBN 3-540-29006-0

Vol. 193. Ingo Glöckner
Fuzzy Quantifiers: A Computational Theory,
2006
ISBN 3-540-29634-4

Vol. 194. Dawn E. Holmes, Lakhmi C. Jain
(Eds.)
Innovations in Machine Learning, 2006
ISBN 3-540-30609-9

Dawn E. Holmes
Lakhmi C. Jain
(Eds.)

Innovations
in Machine Learning

Theory and Applications

 Springer

Professor Dawn E. Holmes
Department of Statistics
and Applied Probability
University of California
at Santa Barbara
South Hall
Santa Barbara, CA 93106-3110
USA
E-mail: holmes@pstat.ucsb.edu

Professor Lakhmi C. Jain
School of Electrical
& Information Engineering
Knowledge-Based
Intelligent Engineering
Mawson Lakes, SA
Adelaide 5095
Australia
E-mail: lakhmi.jain@unisa.edu.au

ISSN print edition: 1434-9922
ISSN electronic edition: 1860-0808

ISBN:13 978-3-642-06788-4
e-ISBN:13 978-3-540-33486-6

Springer is a part of Springer Science+Business Media
springer.com
© Springer-Verlag Berlin Heidelberg 2006
Softcover reprint of the hardcover 1st edition 2006

Cover design: Erich Kirchner, Heidelberg

This book is dedicated to our students.

Foreword

The study of innovation – the development of new knowledge and artifacts – is of interest to scientists and practitioners. Innovations change the day-to-day lives of individuals, transform economies, and even create new societies. The conditions triggering innovation and the innovation process itself set the stage for economic growth.

Scholars of technology have indicated that innovation lies at the intersection of science and technology. One view proposes that innovation is possible through advances in basic science and is realized in concrete products within the context of applied science. Another view states that the development of innovative products through applied science generates new resources on which basic science draws to advance new ideas and theories. Some believe that that science and technology form a symbiotic relationship, drawing from and contributing to one another's progress. Following this view, innovation in any domain can be enhanced by principles and insights from diverse disciplines.

This book addresses an important component of innovation dealing with knowledge discovery. The discovery aspect creates a natural bridge between machine learning concepts, models, and algorithms and innovation. In years to come machine learning will mark some of the early fundamentals leading to innovative science.

Andrew Kusiak
Professor of Mechanical and Industrial Engineering
Intelligent Systems Laboratory
The University of Iowa
Iowa City, Iowa
USA

Preface

There are many invaluable books available on machine learning. However, in compiling a volume titled "Innovations in Machine Learning" we wish to introduce some of the latest developments to a broad audience of both specialists and non-specialists in this field.

So, what is machine learning? Machine learning is a branch of artificial intelligence that grew, as a research area, out of such diverse disciplines as traditional computer science, linguistics, cognitive science, psychology and philosophy. Although the philosophical roots of the subject may be traced back to Leibniz and even ancient Greece, the modern era begins with the work of Norbert Wiener, the father of Cybernetics, a term that he introduced in 'Control and Communication in the Animal and the Machine' (1948). However, it was not until 1955 that 'The Logic Theorist', generally accepted as the first AI program, was presented by Newell and Simon. In this ground-breaking work, Newell and Simon proved that computers were more than just calculating machines, thus shepherding in the era of the computational model of the mind.

In Turing's 1950 seminal work 'Computing Machinery and Intelligence', in which he first presents his famous eponymous test, Turing hoped to establish the claim that human intelligence is not special but can be explained in terms of computation. Research initially focused on the misguided notion that machine intelligence should provide a model for human intelligence. Ultimately, researchers in expert systems found that this was not the way to go.

The intractable question 'Can machines think'? was soon modified to 'Can machines learn'? the answer to which directs the research area that is the subject of this volume. Machine learning is, therefore, concerned with building adaptive computer systems that are capable of improving their performance through learning.

Minsky, a great pioneer in machine learning, built SNARC (Stochastic Neural-Analog Reinforcement computer), the first randomly wired neural network learning machine, in 1951. Machine learning in the 1960's was largely concerned with knowledge representation and heuristic methods but by the early 1970's research in neural networks had begun to flourish. With the fast moving pace of AI research in the 1980's it became realistic to develop systems to solve real-world problems.

As we shall see below, each of the three main learning systems; Symbolic Learning, Neural Networks and Genetic Algorithms are represented in this volume. The pioneering work of Mitchell on Version Spaces resulted in a new paradigm for symbolic inductive learning, which became a dynamic research area. Research in Neural Networks blossomed after the publication of 'Parallel Distributed Processing, Volume I and II' [Rumelhart and James McClelland 1984].

John Holland introduced Genetic Algorithms in the early 1970's. The on-going development of this area has resulted in a major paradigm for research into automated computer program generation. The current volume introduces the reader to research in classifier systems, the area of genetic programming concerned with machine learning.

In compiling this volume we have sought to present innovative research from prestigious contributors in the field of machine learning. Each of the 9 chapters is self-contained and is described below.

Chapter 1 by D. Heckerman C. Meek and G. Cooper is on a Bayesian approach to casual discovery. In addition to describing the general Bayesian approach to causal discovery, the authors present a review of approximation methods for missing data and hidden variables, and illustrate differences between the Bayesian and constraint-based methods using artificial and real examples.

Chapter 2 by Neapolitan and Jiang presents a tutorial on learning casual influences. In the last decade related research in AI, cognitive science and philosophy have resulted in a method for learning casual relationships when we have data on at least four variables. This method is described in this chapter. The recent research on learning casual relationships in the presence of only two variables is also presented.

Chapter 3 by Roth is on learning based programming. The author has proposed a programming model that supports interaction with domain elements at a semantic level. The author has presented some of the theoretical foundations and first generation implementations of the learning based programming language.

In Chapter 4, Eberhardt, Glymour and Scheines have presented their research on casual relations. By combining experimental interventions with search procedures for graphical causal methods, useful relationships are shown with perfect data. This research provides useful insight in active learning scenario.

Chapter 5 by S.H. Muggleton, H. Lodhi, A. Amini and M.J.E. Sternberg is on Support Vector Inductive Logic Programming (SVILP). The authors have proposed a general method for constructing kernels for support vector inductive logic programming. SVIPL is evaluated empirically against related approaches. The experimental results demonstrate that this novel approach significantly outperforms all other approaches in the study.

Chapter 6 by Yoshua Bengio, Holger Schwenk, Jean-Sébastien Senécal, Fréderic Morin and Jean-Luc Gauvain is on neural probabilistic language models. The main goal of statistical language modeling is to learn the joint probability function of sequences of words in a language. The authors have proposed a new scheme to overcome the curse of dimensionality by learning a distributed representation for words. A number of methods are proposed to speed-up both training and probability computation. The authors have incorporated their new model into a speech recognize of conversational speech.

Chapter 7 by Adriaans and van Zaanen is on computational grammatical inference. The authors have presented the overview of this area of research. The authors present linguistic, empirical, and formal grammatical inference and discuss the work that falls in the areas where these fields overlap.

Chapter 8 by Jaz Kandola, John Shawe-Taylor, Andre Elisseeff and Nello Cristianini is on kernel target alignment. Kernal based methods are increasing being used for data modeling. The authors have presented their research on measuring the degree of agreement between a kernel and a learning task.

Chapter 9 by Ralf Herbrich, Thore Graepel, and Bob Williamson is on the structure of version space. The authors have presented their research on generalization performance of consistent classifiers, i.e. classifiers that are contained in the so-called version space. Using a recent result in the PAC-Bayesian framework the authors have shown that given a suitably chose hypothesis space these exists a large fraction of classifiers with small generalization error. The findings are validated using the kernel Gibbs sampler algorithm.

This book will prove valuable to theoreticians as well as application scientists/engineers in the broad area of artificial intelligence. Postgraduate students will also find this a useful sourcebook since it shows the direction of current research.

We have been fortunate in attracting contributions from top class researchers and wish to offer our thanks for their support in this project. We also acknowledge the expertise and time of the reviewers. We appreciate the assistance of Berend Jan van der Zwaag during the final preparation of manuscript. Finally, we also wish to thank Springer-Verlag for their support.

USA   *Dawn E. Holmes*
January 2006   *Lakhmi C. Jain*

Table of Contents

1   A Bayesian Approach to Causal Discovery

David Heckerman[1], Christopher Meek[1] and Gregory Cooper[2]

1. Microsoft Research , Redmond WA, 98052-6399
   heckerma@microsoft.com; meek@microsoft.com
2. University of Pittsburgh, Pittsburgh, PA.
   gfc@smi.med.pitt.edu

Abstract

We examine the Bayesian approach to the discovery of causal DAG models and compare it to the constraint-based approach. Both approaches rely on the Causal Markov condition, but the two differ significantly in theory and practice. An important difference between the approaches is that the constraint-based approach uses categorical information about conditional-independence constraints in the domain, whereas the Bayesian approach weighs the degree to which such constraints hold. As a result, the Bayesian approach has three distinct advantages over its constraint-based counterpart. One, conclusions derived from the Bayesian approach are not susceptible to incorrect categorical decisions about independence facts that can occur with data sets of finite size. Two, using the Bayesian approach, finer distinctions among model structures—both quantitative and qualitative—can be made. Three, information from several models can be combined to make better inferences and to better account for modeling uncertainty. In addition to describing the general Bayesian approach to causal discovery, we review approximation methods for missing data and hidden variables, and illustrate differences between the Bayesian and constraint-based methods using artificial and real examples.

1.1 Introduction

In this paper, we examine the Bayesian approach to the discovery of causal models in the family of directed acyclic graphs (DAGs). The Bayesian approach is related to the constraint-based approach, which is discussed in Chapters 1, 5, and 6 of Heckerman (1996). In particular, both methods rely

D. Heckerman et al.: *A Bayesian Approach to Causal Discovery*, StudFuzz **194**, 1–28 (2006)
www.springerlink.com   © Springer-Verlag Berlin Heidelberg 2006

on the Causal Markov condition. Nonetheless, the two approaches differ significantly in theory and practice. An important difference between them is that the constraint-based approach uses categorical information about conditional-independence constraints in the domain, whereas the Bayesian approach weighs the degree to which such constraints hold. As a result, the Bayesian approach has three distinct advantages over its constraint-based counterpart. One, conclusions derived from the Bayesian approach are not susceptible to incorrect categorical decisions about independence facts that can occur with data sets of finite size. Two, using the Bayesian approach, finer distinctions among model structures—both quantitative and qualitative—can be made. Three, information from several models can be combined to make better inferences and to better account for modeling uncertainty.

In Sections 1.2 and 1.3, we review the Bayesian approach to model averaging and model selection and its application to the discovery of causal DAG models. In Section 1.4, we discuss methods for assigning priors to model structures and their parameters. In Section 1.5, we compare the Bayesian and constraint-based methods for causal discovery for a small domain with complete data, highlighting some of the advantages of the Bayesian approach. In Section 1.6, we note computational difficulties associated with the Bayesian approach when data sets are incomplete—for example, when some variables are hidden—and discuss more efficient approximation methods including Monte-Carlo and asymptotic approximations. In Section 1.7, we illustrate the Bayesian approach on the data set of Sewall and Shah (1968) concerning the college plans of high-school students. Using this example, we show that the Bayesian approach can make finer distinctions among model structures than can the constraint-based approach.

1.2 The Bayesian Approach

In a constraint-based approach to the discovery of causal DAG models, we use data to make *categorical* decisions about whether or not particular conditional-independence constraints hold. We then piece these decisions together by looking for those sets of causal structures that are consistent with the constraints. To do so, we use the Causal Markov condition (Spirtes et. al, 1993) to link lack of cause with conditional independence.

In the Bayesian approach, we also use the Causal Markov condition to look for structures that fit conditional-independence constraints. In contrast to constraint-based methods, however, we use data to make probabilistic inferences about conditional-independence constraints. For example, rather than conclude categorically that, given data, variables X and Y are independent, we conclude that these variables are independent with some probability. This probability encodes our uncertainty about the presence or absence of independence. Furthermore, because the Bayesian approach uses a probabilistic framework, we no longer need to make decisions about individual independence facts. Rather, we compute the probability that the independencies associated with an entire causal structure are true. Then, using such probabilities, we can average a particular hypothesis of interest, such as "Does X cause Y?" over all possible causal structures.

Let us examine the Bayesian approach in some detail. Suppose our problem domain consists of variables $\mathbf{X} = \{X_1,...,X_n\}$. In addition, suppose that we have some data $D = \{\mathbf{x}_1,...,\mathbf{x}_N\}$, which is a random sample from some unknown probability distribution for $\mathbf{X}$. For the moment, we assume that each case $\mathbf{x}$ in D consists of an observation of all the variables in $\mathbf{X}$. We assume that the unknown probability distribution can be encoded by some causal model with structure $\mathbf{m}$. As in Spirtes *et al* (1993), we assume that the structure of this causal model is a DAG that encodes conditional independencies via the Causal Markov condition. We are uncertain about the structure and parameters of the model; and—using the Bayesian approach—we encode this uncertainty using probability. In particular, we define a discrete variable $\mathbf{M}$ whose states $\mathbf{m}$ correspond to the possible true models, and encode our uncertainty about $\mathbf{M}$ with the probability distribution $p(\mathbf{m})$. In addition, for each model structure $\mathbf{m}$, we define a continuous vector-valued variable Θ_m, whose values θ_m correspond to the possible true parameters. We encode our uncertainty about Θ_m using the (smooth) probability density function $p(\theta_m | \mathbf{m})$. The assumption that $p(\theta_m | \mathbf{m})$ is a smooth probability density function entails the assumption of faithfulness employed in constraint-based methods for causal discovery (Meek, 1995).

Given random sample D, we compute the posterior distributions for each $\mathbf{m}$ and θ_m using Bayes' rule:

$$p(\mathbf{m} | D) = \frac{p(\mathbf{m})p(D | \mathbf{m})}{\sum_{m'} p(\mathbf{m}')p(D | \mathbf{m}')}$$

$$p(\theta_m | D, \mathbf{m}) = \frac{p(\theta_m | \mathbf{m})p(D | \theta_m, \mathbf{m})}{p(D | \mathbf{m})} \tag{2}$$

where

$$p(D\,|\,\mathbf{m}) = \int p(D\,|\,\theta_m,\mathbf{m})p(\theta_m\,|\,\mathbf{m})d\theta_m \qquad (3)$$

is called the *marginal likelihood*. Given some hypothesis of interest, h, we determine the probability that h is true given data D by averaging over all possible models and their parameters:

$$p(h\,|\,D) = \sum_m p(\mathbf{m}\,|\,D)p(h\,|\,D,\mathbf{m}) \qquad (4)$$

$$p(h\,|\,D,\mathbf{m}) = \int p(h\,|\,\theta_m,\mathbf{m})p(\theta_m\,|\,D,\mathbf{m})d\theta_m \qquad (5)$$

For example, h may be the event that the next case $\mathbf{X}_{N+1}$ is observed in configuration $\mathbf{x}_{N+1}$. In this situation, we obtain

$$p(\mathbf{x}_{N+1}\,|\,D) = \sum_m p(\mathbf{m}\,|\,D)\int p(\mathbf{x}_{N+1}\,|\,\theta_m,\mathbf{m})p(\theta_m\,|\,D,\mathbf{m})d\theta_m \qquad (6)$$

where $p(\mathbf{x}_{N+1}\,|\,\theta_m,\mathbf{m})$ is the likelihood for the model. As another example, h may be the hypothesis that "X causes Y". We consider such a situation in detail in Section 1.5.

Under certain assumptions, these computations can be done efficiently and in closed form. One assumption is that the likelihood term $p(x\,|\,\theta_m,\mathbf{m})$ factors as follows:

$$p(\mathbf{x}\,|\,\theta_m,\mathbf{m}) = \prod_{i=1}^n p(x_i\,|\,\mathbf{pa}_i,\theta_i,\mathbf{m}) \qquad (7)$$

where each *local likelihood* $p(x_i\,|\,\mathbf{pa}_i,\theta_i,\mathbf{m})$ is in the exponential family. In this expression, $\mathbf{pa}_i$ denotes the configuration of the variables corresponding to parents of node x_i, and θ_i denotes the set of parameters associated with the local likelihood for variable x_i. One example of such a factorization occurs when each variable $X_i \in \mathbf{X}$ is discrete, having r_i possible values $x_i^1,...,x_i^{r_i}$, and each local likelihood is a collection of multinomial distributions, one distribution for each configuration of $\mathbf{Pa}_i$, that is,

$$p(x_i^k\,|\,\mathbf{pa}_i^j,\theta_i,\mathbf{m}) = \theta_{ijk} > 0 \qquad (8)$$

where $\mathbf{pa}_i^1,...,\mathbf{pa}_i^{q_i}$ ($q_i = \prod_{X_i \in \mathbf{Pa}_i} r_i$) denote the configurations of $\mathbf{Pa}_i$, and $\theta_i = ((\theta_{ijk})_{k=2}^{r_i})_{j=1}^{q_i}$ are the parameters. The parameter θ_{ij1} is given by

$1 - \sum_{k=2}^{r_i} \theta_{ijk}$. We shall use this example to illustrate many of the concepts in this paper. For convenience, we define the vector of parameters

$$\theta_{ij} = (\theta_{ij2}, ..., \theta_{ijr_i})$$

for all i and j. A second assumption for efficient computation is that the parameters are mutually independent. For example, given the discrete-multinomial likelihoods, we assume that the parameter vectors θ_{ij} are mutually independent.

Let us examine the consequences of these assumptions for our multinomial example. Given a random sample D that contains no missing observations, the parameters remain independent:

$$p(\theta_m \mid D, \mathbf{m}) = \prod_{i=1}^{n} \prod_{j=1}^{q_i} p(\theta_{ij} \mid D, \mathbf{m}) \tag{9}$$

Thus, we can update each vector of parameters θ_{ij} independently. Assuming each vector θ_{ij} has a conjugate prior[1], namely, a Dirichlet distribution $\mathrm{Dir}(\theta_{ij} \mid \alpha_{ij1}, ..., \alpha_{ijr_i})$ —we obtain the posterior distribution for the parameters

$$p(\theta_{ij} \mid D, \mathbf{m}) = \mathrm{Dir}(\theta_{ij} \mid \alpha_{ij1} + N_{ij1}, ..., \alpha_{ijr_i} + N_{ijr_i}) \tag{10}$$

where N_{ijk} is the number of cases in D in which $X_i = x_i^k$ and $\mathbf{Pa}_i = \mathbf{pa}_i^j$. Note that the collection of counts N_{ijk} are sufficient statistics of the data for the model $\mathbf{m}$. In addition, we obtain the marginal likelihood (derived in Cooper and Herskovits, 1992):

$$p(D \mid \mathbf{m}) = \prod_{i=1}^{n} \prod_{j=1}^{q_i} \frac{\Gamma(\alpha_{ij})}{\Gamma(\alpha_{ij} + N_{ij})} \cdot \prod_{k=1}^{r_i} \frac{\Gamma(\alpha_{ijk} + N_{ijk})}{\Gamma(\alpha_{ijk})} \tag{11}$$

where $\alpha_{ij} = \sum_{k=1}^{r_i} \alpha_{ijk}$ and $N_{ij} = \sum_{k=1}^{r_i} N_{ijk}$. We then use Equation 1 and Equation 11 to compute the posterior probabilities $p(\mathbf{m} \mid D)$.

As a simple illustration of these ideas, suppose our hypothesis of interest is the outcome of $\mathbf{X}_{N+1}$, the next case to be seen after D. Also suppose that, for each possible outcome $\mathbf{x}_{N+1}$ of $\mathbf{X}_{N+1}$, the value of X_i is x_i^k and the configuration of $\mathbf{Pa}_i$ is $\mathbf{pa}_i^j$, where k and j depend on i. To

---

[1]Bernardo and Smith (1994) provide a summary of likelihoods from the exponential family and their conjugate priors.

compute $p(\mathbf{x}_{N+1} \mid D)$, we first average over our uncertainty about the parameters. Using Equations 4, 7, and 8, we obtain

$$p(\mathbf{x}_{N+1} \mid D, \mathbf{m}) = \int \left(\prod_{i=1}^{n} \theta_{ijk} \right) p(\theta_m \mid D, \mathbf{m}) d\theta_m$$

Because parameters remain independent given D, we get

$$p(\mathbf{x}_{N+1} \mid D, \mathbf{m}) = \prod_{i=1}^{n} \int p(\theta_{ij} \mid D, \mathbf{m}) \, d\theta_{ij}$$

Because each integral in this product is the expectation of a Dirichlet distribution, we have

$$p(\mathbf{x}_{N+1} \mid D, \mathbf{m}) = \prod_{i=1}^{n} \frac{\alpha_{ijk} + N_{ijk}}{\alpha_{ij} + N_{ij}} \tag{12}$$

Finally, we average this expression for $p(\mathbf{x}_{N+1} \mid D, \mathbf{m})$ over the possible models using Equation 5 to obtain $p(\mathbf{x}_{N+1} \mid D)$.

1.3 Model Selection and Search

The full Bayesian approach is often impractical, even under the simplifying assumptions that we have described. One computation bottleneck in the full Bayesian approach is averaging over all models in Equation 4. If we consider causal models with n variables, the number of possible structure hypotheses is at least exponential in n. Consequently, in situations where we can not exclude almost all of these hypotheses, the approach is intractable. Statisticians, who have been confronted by this problem for decades in the context of other types of models, use two approaches to address this problem: *model selection* and *selective model averaging*. The former approach is to select a "good" model (i.e., structure hypothesis) from among all possible models, and use that model as if it were the correct model. The latter approach is to select a manageable number of good models from among all possible models and pretend that these models are exhaustive. These related approaches raise several important questions. In particular, do these approaches yield accurate results when applied to causal structures? If so, how do we search for good models?

The question of accuracy is difficult to answer in theory. Nonetheless, several researchers have shown experimentally that the selection of a single model that is likely a posteriori often yields accurate predictions (Cooper and Herskovits 1992; Aliferis and Cooper 1994; Heckerman *et al*, 1995) and that selective model averaging using Monte-Carlo methods can sometimes be efficient and yield even better predictions (Herskovits 1991); Madigan *et al*, (1996).

Chickering (1996a) has shown that for certain classes of prior distributions the problem of finding the model with the highest posterior is NP-Complete. However, a number of researchers have demonstrated that greedy search methods over a search space of DAGs works well. Also, constraint-based methods have been used as a first-step heuristic search for the most likely causal model (Singh and Valtorta, 1993; Spirtes and Meek, 1995). In addition, performing greedy searches in a space where Markov equivalent models (see definition below) are represented by a single model has improved performance (Spirtes and Meek, 1995; Chickering 1996b).

1.4 Priors

To compute the relative posterior probability of a model structure, we must assess the structure prior $p(\mathbf{m})$ and the parameter priors $p(\theta_m \mid \mathbf{m})$. Unfortunately, when many model structures are possible, these assessments will be intractable. Nonetheless, under certain assumptions, we can derive the structure and parameter priors for many model structures from a manageable number of direct assessments.

1.4.1 Priors for Model Parameters

First, let us consider the assessment of priors for the parameters of model parameters. We consider the approach of Heckerman *et al.* (1995) who address the case where the local likelihoods are multinomial distributions and the assumption of parameter independence holds.

Their approach is based on two key concepts: *Markov equivalence* and distribution equivalence. We say that two model structures for $\mathbf{X}$ are Markov equivalent if they represent the same set of conditional-independence assertions for $\mathbf{X}$ (Verma and Pearl, 1990). For example, given $\mathbf{X} = \{X, Y, Z\}$, the model structures $X \to Y \to Z$, $X \leftarrow Y \to Z$, and $X \leftarrow Y \leftarrow Z$ represent only the independence assertion that X and Z are conditionally independent given Y. Consequently, these model structures are equivalent. Another example of Markov equivalence is the set of *complete model structures* on $\mathbf{X}$; a complete model is one that has no missing edge and which encodes no assertion of conditional independence. When $\mathbf{X}$ contains n variables, there are $n!$ possible complete model structures; one model structure for each possible ordering of the variables. All complete model structures for $p(\mathbf{X})$ are Markov equivalent. In general, two model structures are Markov equivalent if and only if they have the same structure ignoring arc directions and the same v-structures (Verma and Pearl, 1990). A v-structure is an ordered tuple (X, Y, Z) such that there is an arc from X to Y and from Z to Y, but no arc between X and Z.

The concept of distribution equivalence is closely related to that of Markov equivalence. Suppose that all causal models for $\mathbf{X}$ under consideration have local likelihoods in the family F. This is not a restriction, per se, because F can be a large family. We say that two model structures m_1 and m_2 for $\mathbf{X}$ are *distribution equivalent with respect to (wrt)* F if they represent the same joint probability distributions for $\mathbf{X}$—that is, if, for every θ_{m_1}, there exists a θ_{m_2} such that $p(\mathbf{x} \mid \theta_{m_1}, m_1) = p(\mathbf{x} \mid \theta_{m_2}, m_2)$, and vice versa.

Distribution equivalence wrt some F implies Markov equivalence, but the converse does not hold. For example, when F is the family of logistic-regression models, the complete model structures for $n \geq 3$ variables do not represent the same sets of distributions. Nonetheless, there are families F —for example, multinomial distributions and linear-regression models with Gaussian noise—where Markov equivalence implies distribution equivalence wrt F (Heckerman and Geiger, 1996). The notion of distribution equivalence is important, because if two model structures 1 and 2 are distribution equivalent wrt to a given F, then it is often reasonable to expect that data can not help to discriminate them. That is, we expect $p(D \mid m_1) = p(D \mid m_2)$ for any data set D. Heckerman *et al.* (1995) call this property *likelihood equivalence*. Note that the constraint-based approach also does not discriminate among Markov equivalent structures.

Now let us return to the main issue of this section: the derivation of priors from a manageable number of assessments. Geiger and Heckerman (1995) show that the assumptions of parameter independence and likelihood equivalence imply that the parameters for any *complete* model structure m_c must have a Dirichlet distribution with constraints on the hyperparameters given by equation

$$\alpha_{ijk} = \alpha p(x_i^k, \mathbf{pa}_i^j \mid m_c) \qquad (13)$$

where α is the user's equivalent sample size and $p(x_i^k, \mathbf{pa}_i^j \mid m_c)$ is computed from the user's joint probability distribution $p(\mathbf{x} \mid m_c)$. This result is rather remarkable, as the two assumptions leading to the constrained Dirichlet solution are qualitative.

To determine the priors for parameters of *incomplete* model structures, (Heckerman *et al*, 1995) use the assumption of *parameter modularity*, which says that if X_i has the same parents in model structures m_1 and m_2 then

$$p(\theta_{ij} \mid m_1) = p(\theta_{ij} \mid m_2)$$

for $j = 1, ..., q_i$. They call this property parameter modularity, because it says that the distributions for parameters θ_{ij} depend only on the structure of the model that is local to variable X_i, namely X_i and its parents.

Given the assumptions of parameter modularity and parameter independence, it is a simple matter to construct priors for the parameters of an arbitrary model structure given the priors on complete model structures. In particular, given parameter independence, we construct the priors for the parameters of each node separately. Furthermore, if node X_i has parents $\mathbf{Pa}_i$ in the given model structure, we identify a complete model structure where X_i has these parents, and use Equation 13 and parameter modularity to determine the priors for this node. The result is that all terms α_{ijk} for all model structures are determined by Equation 13. Thus, from the assessments α and $p(\mathbf{x} \mid m_c)$, we can derive the parameter priors for all possible model structures. We can assess $p(\mathbf{x} \mid m_c)$ by constructing a causal model called a prior model that encodes this joint distribution. Heckerman *et al.* (1995) discuss the construction of this model.

1.4.2 Priors for Model Structures

Now, let us consider the assessment of priors on model structures. The simplest approach for assigning priors to model structures is to assume that every structure is equally likely. Of course, this assumption is typically inaccurate and used only for the sake of convenience. A simple refinement of this approach is to ask the user to exclude various structures (perhaps based on judgments of cause and effect) and then impose a uniform prior on the remaining structures. We illustrate this approach in Section 1.7.

Buntine (1991) describes a set of assumptions that leads to a richer yet efficient approach for assigning priors. The first assumption is that the variables can be ordered, (e.g. through a knowledge of time precedence). The second assumption is that the presence or absence of possible arcs are mutually independent. Given these assumptions, $n(n-1)/2$ probability assessments, one for each possible arc in an ordering, determines the prior probability of every possible model structures. One extension to this approach is to allow for multiple possible orderings. One simplification is to assume that the probability that an arc is absent or present is independent of the specific arc in question. In this case, only one probability assessment is required.

An alternative approach, described by Heckerman *et al.* (1995) uses a prior model. The basic idea is to penalize the prior probability of any structure according to some measure of deviation between that structure and the prior model. Heckerman *et al.* (1995) suggest one reasonable measure of deviation.

Madigan *et al.* (1995) give yet another approach that makes use of imaginary data from a domain expert. In their approach, a computer program helps the user create a hypothetical set of complete data. Then, using techniques such as those in Section 1.2 they compute the posterior probabilities of model structures given this data, assuming the prior probabilities of structures are uniform. Finally, they use these posterior probabilities as priors for the analysis of the real data.

1.5 Example

In this section, we provide a simple example that applies Bayesian model averaging and Bayesian model selection to the problem of causal discovery. In addition, we compare these methods with a constraint-based approach.

Let us consider a simple domain containing three binary variables X, Y, and Z. Let h denote the hypothesis that variable X causally influences variable Z. For brevity, we will sometimes state h as "X causes Z".

First, let us consider Bayesian model averaging. In this approach, we use Equation 4 to compute the probability that h is true given data D. Because our models are causal, the expression $p(D \mid \mathbf{m})$ reduces to an index function that is true when $\mathbf{m}$ contains an arc from node X to node Z. Thus, the right-hand-side of Equation 4 reduces to $\sum_{m''} p(\mathbf{m}'' \mid D)$, where the sum is taken over all causal models $\mathbf{m}''$ that contain an arc from X to Z. For our three-variable domain, there are 25 possible causal models and, of these, there are eight models containing an arc from X to Z.

To compute $p(\mathbf{m} \mid D)$, we apply Equation 1, where the sum over $\mathbf{m}'$ is taken over the 25 models just mentioned. We assume a uniform prior distribution over the 25 possible models, so that $p(\mathbf{m}') = 1/25$ for every $\mathbf{m}'$. We use Equation 11 to compute the marginal likelihood $p(D \mid \mathbf{m})$. In applying Equation 11, we use the prior given by $\alpha_{ijk} = 1/r_iq_i$, which we obtain from Equation 13 using a uniform distribution for $p(\mathbf{x} \mid \mathbf{m}_c)$ and an equivalent sample $\alpha = 1$. Because this equivalent sample size is small, the data strongly influences the posterior probabilities for h that we derive.

To generate data, we first selected the model structure $X \longrightarrow Z \longleftarrow Y$ and randomly sampled its probabilities from a uniform distribution. The resulting model is shown in Figure 1.

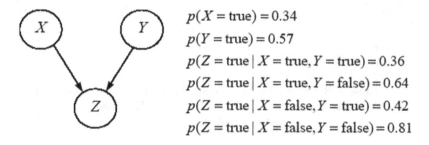

$$p(X = \text{true}) = 0.34$$
$$p(Y = \text{true}) = 0.57$$
$$p(Z = \text{true} \mid X = \text{true}, Y = \text{true}) = 0.36$$
$$p(Z = \text{true} \mid X = \text{true}, Y = \text{false}) = 0.64$$
$$p(Z = \text{true} \mid X = \text{false}, Y = \text{true}) = 0.42$$
$$p(Z = \text{true} \mid X = \text{false}, Y = \text{false}) = 0.81$$

Fig 1: A causal model used to generate data.

Next, we sampled data from the model according to its joint distribution. As we sampled the data, we kept a running total of the number cases seen in each possible configuration of $\{X, Y, Z\}$. These counts are sufficient statistics of the data for any causal model $\mathbf{m}$. These statistics are shown in Table 1 for the first 150, 250, 500, 1000, and 2000 cases in the data set.

number	sufficient statistics						
of cases	$\bar{x}\bar{y}\bar{z}$	$\bar{x}\bar{y}z$	$\bar{x}y\bar{z}$	$\bar{x}yz$	$x\bar{y}\bar{z}$	$x\bar{y}z$	$xy\bar{z}$	xyz
150	5	36	38	15	7	16	23	10
250	10	60	51	27	15	25	41	21
500	23	121	103	67	19	44	79	44
1000	44	242	222	152	51	80	134	75
2000	88	476	431	311	105	180	264	145

Table 1: A summary of data used in the example

The second column in Table 2 shows the results of applying Equation 4 under the assumptions stated above for the first N cases in the data set. When $N = 0$, the data set is empty, in which case probability of hypothesis h is just the prior probability of "X causes Z": 8/25=0.32. Table 2 shows that as the number of cases in the database increases, the probability that "X causes Z" increases monotonically as the number of cases increases. Although not shown, the probability increases toward 1 as the number of cases increases beyond 2000.

number of cases	$p($"X causes Z"$\|D)$	output of Bayesian model selection	output of PC algorithm
150	0.036	X and Z unrelated	X and Z unrelated
250	0.123	X and Z unrelated	X causes Z
500	0.141	X causes Z or Z causes X	X and Z unrelated (with inconsistency)
1000	0.593	X causes Z	X causes Z
2000	0.926	X causes Z	X causes Z

Table 2: Bayesian model averaging, Bayesian model selection and constraint-based results for an analysis of whether "X causes Z" given data summarized in Table 1.

Column 3 in Table 2 shows the results of applying Bayesian model selection. Here, we list the causal relationship(s) between X and Z found in the model or models with the highest posterior probability $p(\mathbf{m}\,|\,D)$. For example, when $N = 500$, there are three models that have the highest posterior probability. Two of the models have Z as a cause of X; and one has X as a cause of Z.

Column 4 in Table 2 shows the results of applying the PC constraint-based causal discovery algorithm (Spirtes *et al*, 1993), which is part of the Tetrad II system (Scheines *et al*, 1994). PC is designed to discover causal relationships that are expressed using DAGs.[2] We applied PC using its default settings, which include a statistical significance level of 0.05. Note that, for $N = 500$, the PC algorithm detected an inconsistency. In particular, the independence tests yielded (1) X and Z are dependent, (2) Y and Z are dependent, (3) X and Y are independent given Z, and (4) X and Z are independent given Y. These relationships are not consistent with the assumption underlying the PC algorithm that the only

---

[2]The algorithm assumes that there are no hidden variables. See Sections 6 and 7 for a discussion of hidden-variable models and methods for learning them.

independence facts found to hold in the sample are those entailed by the Causal Markov condition applied to the generating model. In general, inconsistencies may arise due to the use of thresholds in the independence tests.

There are several weaknesses of the Bayesian-model-selection and constraint-based approaches illustrated by our results. One is that the output is categorical—there is no indication of the strength of the conclusion. Another is that the conclusions may be incorrect in that they disagree with the generative model. Model averaging (column 2) does not suffer from these weaknesses, because it indicates the strength of a causal hypothesis.

Although not illustrated here, another weakness of constraint-based approaches is that their output depends on the threshold used in independence tests. For causal conclusions to be correct asymptotically, the threshold must be adjusted as a function of sample size (N). In practice, however, it is unclear what this function should be.

Finally, we note that there are practical problems with model averaging. In particular, the domain can be so large that there are too many models over which to average. In such situations, the exact probabilities of causal hypotheses can not be calculated. However, we can use selective model averaging to derive approximate posterior probabilities, and consequently give some indication of the strength of causal hypotheses.

1.6 Methods for Incomplete Data and Hidden Variables

Among the assumptions that we described in Section 1.2, the one that is most often violated is the assumption that all variables are observed in every case. In this section, we examine Bayesian methods for relaxing this assumption.

An important distinction for this discussion is that of hidden versus observable variable. A *hidden variable* is one that is unknown in all cases. An *observable variable* is one that is known in some (but not necessarily all) of the cases. We note that constraint-based and Bayesian methods differ significantly in the way that they missing data. Whereas constraint-based methods typically throw out cases that contain an observable variable with a missing value, Bayesian methods do not.

Another important distinction concerning missing data is whether or not the absence of an observation is dependent on the actual states of the variables. For example, a missing datum in a drug study may indicate that a patient became too sick—perhaps due to the side effects of the drug—to continue in the study. In contrast, if a variable is hidden, then the absence of this data is independent of state. Although Bayesian methods and graphical models are suited to the analysis of both situations, methods for handling missing data where absence is independent of state are simpler than those where absence and state are dependent. Here, we concentrate on the simpler situation. Readers interested in the more complicated case should see Rubin (1978), Robins (1986), Cooper (1995), and Spirtes *et al.* (1995).

Continuing with our example using discrete-multinomial likelihoods, suppose we observe a single incomplete case. Let $\mathbf{Y} \subset \mathbf{X}$ and $\mathbf{Z} = \mathbf{X} \setminus \mathbf{Y}$ denote the observed and unobserved variables in the case, respectively. Under the assumption of parameter independence, we can compute the posterior distribution of θ_{ij} for model structure $\mathbf{m}$ as follows:

$$p(\theta_{ij} \mid \mathbf{y}, \mathbf{m}) = \sum_{\mathbf{z}} p(\mathbf{z} \mid \mathbf{y}, \mathbf{m}) p(\theta_{ij} \mid \mathbf{y}, \mathbf{z}, \mathbf{m})$$

$$= (1 - p(\mathbf{pa}_i^j \mid \mathbf{y}, \mathbf{m}))\{p(\theta_{ij} \mid \mathbf{m})\} + \sum_{k=1}^{r_i} p(x_i^k, \mathbf{pa}_i^j \mid \mathbf{y}, \mathbf{m}) p(\theta_{ij} \mid x_i^k, \mathbf{pa}_i^j, \mathbf{m})$$

$$(14)$$

(See Spiegelhalter and Lauritzen, 1990, for a derivation.) Each term $p(\theta_{ij} \mid x_i^k, \mathbf{pa}_i^j, \mathbf{m})$ in Equation 13 is a Dirichlet distribution. Thus, unless both X_i and all the variables in $\mathbf{Pa}_i$ are observed in case $\mathbf{y}$, the posterior distribution of θ_{ij} will be a linear combination of Dirichlet distributions; that is, a Dirichlet mixture with mixing coefficients $(1 - p(\mathbf{pa}_i^j \mid \mathbf{y}, \mathbf{m}))$ and $p(x_i^k, \mathbf{pa}_i^j \mid \mathbf{y}, \mathbf{m}), k = 1, ..., r_i$.

When we observe a second incomplete case, some or all of the Dirichlet components in Equation 13 will again split into Dirichlet mixtures. That is, the posterior distribution for θ_{ij} will become a mixture of Dirichlet mixtures. As we continue to observe incomplete cases, each missing values for $\mathbf{Z}$, the posterior distribution for θ_{ij} will contain a number of components that is exponential in the number of cases. In general, for any interesting set of local likelihoods and priors, the exact computation of the posterior distribution for θ_m will be intractable. Thus, we require an approximation for incomplete data.

1.6.1 Monte-Carlo Method

One class of approximations is based on Monte-Carlo or sampling methods. These approximations can be extremely accurate, provided one is willing to wait long enough for the computations to converge.

In this section, we discuss one of many Monte-Carlo methods known as *Gibbs sampling*, introduced by Geman and Geman (1984). Given variables $\mathbf{X} = \{X_1,...,X_n\}$ with some joint distribution $p(\mathbf{x})$, we can use a Gibbs sampler to approximate the expectation of a function $f(\mathbf{x})$ with respect to $p(\mathbf{x})$ as follows. First, we choose an initial state for each of the variables in $\mathbf{X}$ somehow (e.g., at random). Next, we pick some variable X_i, unassign its current state, and compute its probability distribution given the states of the other $n-1$ variables. Then, we sample a state for X_i based on this probability distribution, and compute $f(\mathbf{x})$. Finally, we iterate the previous two steps, keeping track of the average value of $f(\mathbf{x})$. In the limit, as the number of cases approach infinity, this average is equal to $E_{p(\mathbf{x})}(f(\mathbf{x}))$ provided two conditions are met. First, the Gibbs sampler must be *irreducible*. That is, the probability distribution $p(\mathbf{x})$ must be such that we can eventually sample any possible configuration of $\mathbf{X}$ given any possible initial configuration of $\mathbf{X}$. For example, if $p(\mathbf{x})$ contains no zero probabilities, then the Gibbs sampler will be irreducible. Second, each X_i must be chosen infinitely often. In practice, an algorithm for deterministically rotating through the variables is typically used. Introductions to Gibbs sampling and other Monte-Carlo methods—including methods for initialization and a discussion of convergence—are given by Neal (1993) and Madigan and York (1995).

To illustrate Gibbs sampling, let us approximate the probability density $p(\theta_m \mid D, \mathbf{m})$ for some particular configuration of θ_m, given an incomplete data set $D = \{\mathbf{y}_1,...,\mathbf{y}_N\}$ and a causal model for discrete variables with independent Dirichlet priors. To approximate $p(\theta_m \mid D, \mathbf{m})$, we first initialize the states of the unobserved variables in each case somehow. As a result, we have a complete random sample D_c. Second, we choose some variable X_{il} (variable X_i in case l) that is not observed in the original random sample D, and reassign its state according to the probability distribution

$$p(x'_{il} \mid D_c \setminus x_{il}, \mathbf{m}) = \frac{p(x'_{il}, D_c \setminus x_{il} \mid \mathbf{m})}{\sum_{x''_{il}} p(x''_{il}, D_c \setminus x_{il} \mid \mathbf{m})}$$

where $D_c \setminus x_{il}$ denotes the data set D_c with observation x_{il} removed, and the sum in the denominator runs over all states of variable X_{il}. As we have seen, the terms in the numerator and denominator can be computed efficiently (see Equation 11). Third, we repeat this reassignment for all unobserved variables in D, producing a new complete random sample D'_c. Fourth, we compute the posterior density $p(\theta_m | D'_c, \mathbf{m})$ as described in Equations 9 and 10. Finally, we iterate the previous three steps, and use the average of $p(\theta_m | D'_c, \mathbf{m})$ as our approximation.

Monte-Carlo approximations are also useful for computing the marginal likelihood given incomplete data. One Monte-Carlo approach, described by Chib (1995) and Raftery (1996), uses Bayes' theorem:

$$p(D | \mathbf{m}) = \frac{p(\theta_m | \mathbf{m}) p(D | \theta_m, \mathbf{m})}{p(\theta_m | D, \mathbf{m})} \qquad (15)$$

For any configuration of θ_m, the prior term in the numerator can be evaluated directly. In addition, the likelihood term in the numerator can be computed using causal-model inference (Jensen *et al*, 1990). Finally, the posterior term in the denominator can be computed using Gibbs sampling, as we have just described. Other, more sophisticated Monte-Carlo methods are described by DiCiccio *et al.* (1995).

6.1.2 The Gaussian Approximation

Monte-Carlo methods yield accurate results, but they are often intractable—for example, when the sample size is large. Another approximation that is more efficient than Monte-Carlo methods and often accurate for relatively large samples is the *Gaussian approximation* (e.g., Kass *et al*, 1988; Kass and Raftery, 1995).

The idea behind this approximation is that, for large amounts of data, $p(\theta_m | D, \mathbf{m}) \propto p(D | \theta_m, \mathbf{m}) \cdot p(\theta_m | \mathbf{m})$ can often be approximated as a multivariate-Gaussian distribution. In particular, let

$$g(\theta_m) \equiv \log(p(D | \theta_m, \mathbf{m}) \cdot p(\theta_m | \mathbf{m})) \qquad (16)$$

Also, define θ_m to be the configuration of θ_m that maximizes $g(\theta_m)$. This configuration also maximizes $p(\theta_m | D, \mathbf{m})$, and is known as the *maximum a posteriori* (MAP) configuration of θ_m. Using a second degree Taylor polynomial of $g(\theta_m)$ about the θ_m to approximate $g(\theta_m)$, we obtain

$$g(\theta_m) \approx g(\theta_m) - \frac{1}{2}(\theta_m - \theta_m) A (\theta_m - \theta_m)^t \qquad (17)$$

where $(\theta_m - \theta_m)^t$ is the transpose of row vector $(\theta_m - \theta_m)$, and A is the negative Hessian of $g(\theta_m)$ evaluated at θ_m. Raising $g(\theta_m)$ to the power of e and using Equation 15, we obtain

$$p(\theta_m \mid D, \mathbf{m}) \propto p(D \mid \theta_m, \mathbf{m}) p(\theta_m \mid \mathbf{m})$$

$$\approx p(D \mid \theta_m, \mathbf{m}) p(\theta_m \mid \mathbf{m}) \exp\{-\frac{1}{2}(\theta_m - \theta_m) A (\theta_m - \theta_m)^t\} \ (18)$$

Hence, the approximation for $p(\theta_m \mid D, \mathbf{m})$ is Gaussian.

To compute the Gaussian approximation, we must compute θ_m as well as the negative Hessian of $g(\theta_m)$ evaluated at θ_m. In the following section, we discuss methods for finding θ_m. Meng and Rubin (1991) describe a numerical technique for computing the second derivatives. Raftery (1995) shows how to approximate the Hessian using likelihood-ratio tests that are available in many statistical packages. Thiesson (1995) demonstrates that, for multinomial distributions, the second derivatives can be computed using causal-model inference.

Using the Gaussian approximation, we can also approximate the marginal likelihood. Substituting Equation 17 into Equation 3, integrating, and taking the logarithm of the result, we obtain the approximation:

$$\log p(D \mid \mathbf{m}) \approx \log p(D \mid \theta_m, \mathbf{m}) + \log p(\theta_m \mid \mathbf{m}) + \frac{d}{2}\log(2\pi) - \frac{1}{2}\log|A|$$

$$(18)$$

where d is the dimension of $g(\theta_m)$. For a causal model with multinomial distributions, this dimension is typically given by $\prod_{i=1}^{n} q_i(r_i - 1)$. Sometimes, when there are hidden variables, this dimension is lower. See Geiger et al. (1996) for a discussion of this point. This approximation technique for integration is known as *Laplace's method*, and we refer to Equation 18 as the *Laplace approximation*. Kass et al. (1988) have shown that, under certain regularity conditions, the relative error of this approximation is $O_p(1/N)$, where N is the number of cases in D. Thus, the Laplace approximation can be extremely accurate. For more detailed discussions of this approximation, see—for example—Kass et al. (1988) and Kass and Raftery (1995).

Although Laplace's approximation is efficient relative to Monte-Carlo approaches, the computation of $|A|$ is nevertheless intensive for large-dimension models. One simplification is to approximate $|A|$ using only the diagonal elements of the Hessian A. Although in so doing, we incorrectly impose independencies among the parameters, researchers have shown that the approximation can be accurate in some circumstances (see, e.g., Becker and Le Cun, 1989, and Chickering and Heckerman, 1997). Another efficient variant of Laplace's approximation is described by Cheeseman and Stutz (1995) and Chickering and Heckerman (1997).

We obtain a very efficient (but less accurate) approximation by retaining only those terms in Equation 18 that increase with N : $\log p(D \mid \theta_m, \mathbf{m})$, which increases linearly with N, and $\log |A|$, which increases as $d \log N$. Also, for large N, θ_m can be approximated by θ_m, the maximum likelihood configuration of θ_m (see the following section). Thus, we obtain

$$\log p(D \mid \mathbf{m}) \approx \log p(D \mid \theta_m, \mathbf{m}) - \frac{d}{2} \log N \qquad (19)$$

This approximation is called the *Bayesian information criterion* (BIC). Schwarz (1978) has shown that the relative error of this approximation is $O_p(1)$ for a limited class of models. Haughton (1988) has extended this result to curved exponential models. Kass and Wasserman (1995).

The BIC approximation is interesting in several respects. First, roughly speaking, it does not depend on the prior. Consequently, we can use the approximation without assessing a prior.[3] Second, the approximation is quite intuitive. Namely, it contains a term measuring how well the parameterized model predicts the data ($\log p(D \mid \theta_m, \mathbf{m})$) and a term that punishes the complexity of the model ($d/2 \log N$). Third, the BIC approximation is exactly minus the Minimum Description Length (MDL) criterion described by Rissanen (1987).

1.6.3 The MAP and ML Approximations and the EM Algorithm

As the sample size of the data increases, the Gaussian peak will become sharper, tending to a delta function at the MAP configuration θ_m. In this limit, we can replace the integral over θ_m in Equation 5 with $p(h \mid \theta_m, \mathbf{m})$. A further approximation is based on the observation that, as the sample size increases, the effect of the prior $p(\theta_m \mid \mathbf{m})$ diminishes. Thus, we can approximate θ_m by the *maximum likelihood* (ML) configuration of θ_m :

$$\theta_m = \arg \max_{\theta_m} \left\{ p(D \mid \theta_m, \mathbf{m}) \right\}$$

---

[3] One of the technical assumptions used to derive this approximation is that the prior is bounded and bounded away from zero around θ_m.

One class of techniques for finding a ML or MAP is gradient-based optimization. For example, we can use gradient ascent, where we follow the derivatives of $g(\theta_m)$ or the likelihood $p(D \mid \theta_m, \mathbf{m})$ to a local maximum. Russell *et al.* (1995) and Thiesson (1995) show how to compute the derivatives of the likelihood for a causal model with multinomial distributions. Buntine (1994) discusses the more general case where the likelihood comes from the exponential family. Of course, these gradient-based methods find only local maxima.

Another technique for finding a local ML or MAP is the expectation–maximization (EM) algorithm (Dempster *et al*, 1977). To find a local MAP or ML, we begin by assigning a configuration to θ_m somehow (e.g., at random). Next, we compute the *expected* sufficient statistics for a complete data set, where expectation is taken with respect to the joint distribution for $\mathbf{X}$ conditioned on the assigned configuration of θ_m and the known data D. In our discrete example, we compute

$$E_{p(\mathbf{x} \mid D, \theta_s, \mathbf{m})}(N_{ijk}) = \sum_{l=1}^{N} p(x_i^k, \mathbf{pa}_i^j \mid \mathbf{y}_l, \theta_m, \mathbf{m}) \tag{20}$$

where $\mathbf{y}_l$ is the possibly incomplete l th case in D. When X_i and all the variables in $\mathbf{Pa}_i$ are observed in case $\mathbf{x}_l$, the term for this case requires a trivial computation: it is either zero or one. Otherwise, we can use any causal-model inference algorithm to evaluate the term. This computation is called the *expectation step* of the EM algorithm.

Next, we use the expected sufficient statistics as if they were actual sufficient statistics from a complete random sample D_c. If we are doing an ML calculation, then we determine the configuration of θ_m that maximizes $p(D_c \mid \theta_m, \mathbf{m})$. In our discrete example, we have

$$\theta_{ijk} = \frac{E_{p(\mathbf{x} \mid D, \theta_s, \mathbf{m})}(N_{ijk})}{\sum_{k=1}^{r_i} E_{p(\mathbf{x} \mid D, \theta_s, \mathbf{m})}(N_{ijk})}$$

If we are doing a MAP calculation, then we determine the configuration of θ_m that maximizes $p(\theta_m \mid D_c, \mathbf{m})$. In our discrete example, we have[4]

---

[4]The MAP configuration θ_m depends on the coordinate system in which the parameter variables are expressed. The MAP given here corresponds to the *canonical* coordinate system for the multinomial distribution (see, e.g., Bernardo and Smith, 1994, pp. 199–202).

$$\theta_{ijk} = \frac{\alpha_{ijk} + E_{p(\mathbf{x}|D,\theta_s,\mathbf{m})}(N_{ijk})}{\sum_{k=1}^{r_i}(\alpha_{ijk} + E_{p(\mathbf{x}|D,\theta_s,\mathbf{m})}(N_{ijk}))}$$

This assignment is called the *maximization step* of the EM algorithm. Under certain regularity conditions, iteration of the expectation and maximization steps will converge to a local maximum. The EM algorithm is typically applied when sufficient statistics exist (i.e., when local likelihoods are in the exponential family), although generalizations of the EM algorithm have been used for more complicated local distributions (see, e.g., McLachlan and Krishnan, 1997).

1.7 A Case Study

To further illustrate the Bayesian approach and differences between it and the constraint-based approach, let us consider the following example. Sewell and Shah (1968) investigated factors that influence the intention of high school students to attend college. They measured the following variables for 10,318 Wisconsin high school seniors: *Sex* (SEX): male, female; *Socioeconomic Status* (SES): low, lower middle, upper middle, high; *Intelligence Quotient* (IQ): low, lower middle, upper middle, high; *Parental Encouragement* (PE): low, high; and *College Plans* (CP): yes, no. Our goal here is to understand the causal relationships among these variables.

The data are described by the sufficient statistics in Table 3. Each entry denotes the number of cases in which the five variables take on some particular configuration. The first entry corresponds to the configuration SEX = male, SES = low, IQ = low, PE = low, and *CP* =yes.

Table 3: Sufficient statistics for the Sewall and Shah (1968) study.

4	349	13	64	9	207	33	72	12	126	38	54	10	67	49	43
2	232	27	84	7	201	64	95	12	115	93	92	17	79	119	59
8	166	47	91	6	120	74	110	17	92	148	100	6	42	198	73
4	48	39	57	5	47	123	90	9	41	224	65	8	17	414	54
5	454	9	44	5	312	14	47	8	216	20	35	13	96	28	24
11	285	29	61	19	236	47	88	12	164	62	85	15	113	72	50
7	163	36	72	13	193	75	90	12	174	91	100	20	81	142	77
6	50	36	58	5	70	110	76	12	48	230	81	13	49	360	98

The remaining entries correspond to configurations obtained by cycling through the states of each variable such that the last variable (CP) varies most quickly. Thus, for example, the upper (lower) half of the table corresponds to male (female) students.

First, let us analyze the data under the assumption that there are no hidden variables. To generate priors for model parameters, we use the method described in Section 1.4.1 with an equivalent sample size of 5 and a prior model where $p(\mathbf{x} \mid \mathbf{m}_c)$ is uniform. (The results are not sensitive to the choice of parameter priors. For example, none of the results reported in this section change qualitatively for equivalent sample sizes ranging from 3 to 40.) For structure priors, we assume that all model structures are equally likely, except, on the basis of prior causal knowledge about the domain, we exclude structures where *SEX* and/or *SES* have parents, and/or *CP* have children. Because the data set is complete, we use Equation 11 to compute the posterior probabilities of model structures. The two most likely model structures found after an exhaustive search over all structures are shown in Figure 2. Note that the most likely graph has a posterior probability that is extremely close to one so that model averaging is not necessary.

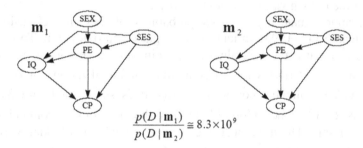

$$\frac{p(D \mid \mathbf{m}_1)}{p(D \mid \mathbf{m}_2)} \cong 8.3 \times 10^9$$

Fig.2: The a posteriori most likely model structures without hidden variables

If we adopt the Causal Markov condition and also assume that there are no hidden variables, then the arcs in both graphs can be interpreted causally. Some results are not surprising—for example the causal influence of socioeconomic status and IQ on college plans. Other results are more interesting. For example, from either graph we conclude that sex influences college plans only indirectly through parental influence. Also, the two graphs differ only by the orientation of the arc between PE and IQ. Either causal relationship is plausible.

We note that the second most likely graph was selected by Spirtes *et al.* (1993), who used the constraint-based PC algorithm with essentially identical assumptions. The only differences in the independence facts entailed by the most likely graph and the second most likely graphs are that the most likely graph entails *SEX* and *IQ* are independent given *SES* and *PE* whereas the second most likely graph entails *SEX* and *IQ* are marginally independent. Although both Bayesian and classical independence tests indicate that the conditional independence holds more strongly given the data, the PC algorithm chooses the second most likely graph due to its greedy nature. In particular, after the PC algorithm decides that *SEX* and *IQ* are marginally independent (at the threshold used by Spirtes *et al*), it never considers the independence of *SEX* and *IQ* given *SES* and *PE* .

Returning to our analysis, the most suspicious result is the suggestion that socioeconomic status has a direct influence on IQ. To question this result, let us consider new models obtained from the models in Figure 2 by replacing this direct influence with a hidden variable pointing to both *SES* and *IQ*. Let us also consider models where (1) the hidden variable points to two or three of *SES*, *IQ*, and *PE* , (2) none, one, or both of the connections *SES — PE* and *PE — IQ* are removed, and (3) no variable has more than three parents. For each structure, we vary the number of states of the hidden variable from two to six.

We approximate the posterior probability of these models using the Cheeseman-Stutz (1995) variant of the Laplace approximation. To find the MAP θ_m, we use the EM algorithm, taking the largest local maximum from among 100 runs with different random initializations of θ_m. The model with the highest posterior probability is shown in Figure 3. This model is $2 \cdot 10^{10}$ times more likely that the best model containing no hidden variable. The next most likely model containing a hidden variable, which has one additional arc from the hidden variable to PE, is $5 \cdot 10^{-9}$ times less likely than the best model. Thus, if we again adopt the Causal Markov condition and also assume that we have not omitted a reasonable model from consideration, then we have strong evidence that a hidden variable is influencing both socioeconomic status and IQ in this population—a sensible result. In particular, according to the probabilities in Figure 3, both *SES* and *IQ* are more likely to be high when H takes on the value 1. This observation suggests that the hidden variable represents "parent quality".

It is possible for constraint-based methods which use independence constraints to discriminate between models with and without hidden variables and to indicate the presence of latent variables (see Spirtes *et al*, 1993). Constraint-based methods also use non-independence constraints;

for example, tetrad constraints, to make additional discriminations. However, these methods cannot distinguish between the model in Figure 3 and the most likely graph without hidden variables (the network on the left in Figure 1). Conditional independence constraints alone cannot be used to distinguish the models, because the two graphs entail the same set of independence facts on the observable variables. Independence constraints in combination with known non-independence constraints also fail to discriminate between the models. In addition, as this study illustrates, Bayesian methods can sometimes be used to determine the number of classes for a latent variable. A constraint-based method using only independence constraints can never determine the number of classes for a latent variable. We conjecture that any distinction among causal structures which can be made by constraint-based methods, even those not restricted to independence constraints, can be made using Bayesian methods. In addition, we conjecture that, asymptotically, when a constraint-based method chooses one model over another, the Bayesian approach will make the same choice, provided the Causal Markov condition and the assumption of faithfulness hold.

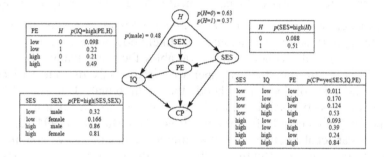

Fig. 3: The a posteriori most likely model structure with a hidden variable. Probabilities shown are MAP values. Some probabilities are omitted through lack of space.

1.8 Open Issues

The Bayesian framework gives us a conceptually simple framework for learning causal models. Nonetheless, the Bayesian solution often comes with a high computational cost. For example, when we learn causal models containing hidden variables, both the exact computation of marginal likelihood and model averaging/selection can be intractable. Although the approximations described in Section 1.6 can be applied to address the difficulties associated with the computation of the marginal likelihood, model averaging and model selection remain difficult.

Another problem associated with learning causal models containing hidden variables is the assessment of parameter priors. The approach in Section 1.4 can be applied in such situations, although the assessment of a joint distribution $p(\mathbf{x} \mid \mathbf{m}_c)$ in which $\mathbf{x}$ includes hidden variables can be difficult. Another approach may be to employ a property called *strong likelihood equivalence* (Heckerman, 1995). According to this property, data should not help to discriminate among two models that are distribution equivalent with respect to the non-hidden variables. Heckerman (1995) showed that any method that uses this property will yield priors that differ from those obtained using a prior network on all variables, both measured and hidden.[5]

One possibility for avoiding this problem with hidden-variable models, when the sample size is sufficiently large, is to use BIC-like approximations. Such approximations are commonly used (Crawford, 1994; Raftery, 1995). Nonetheless, the regularity conditions that guarantee $O_p(1)$ or better accuracy do not typically hold when choosing among causal models with hidden variables. Additional work is needed to obtain accurate approximations for the marginal likelihood of these models.

Even in models without hidden variables there are many interesting issues to be addressed. In this paper we discuss only discrete variables having one type of local likelihood: the multinomial. Thiesson (1995) discusses a class of local likelihoods for discrete variables that use fewer parameters. Geiger and Heckerman (1994) and Buntine (1994) discuss simple linear local likelihoods for continuous nodes that have continuous and discrete variables. Buntine (1994) also discusses a general class of local likelihoods from the exponential family for nodes having no parents. Nonetheless, alternative likelihoods for discrete and continuous variables are desired. Local likelihoods with fewer parameters might allow for the selection of correct models with less data. In addition, local likelihoods that express more accurately the data generating process would allow for easier interpretation of the resulting models.

---

[5]In particular, Heckerman (1995) showed that strong likelihood equivalence is not consistent with parameter independence and parameter modularity.

Acknowledgments

We thank Max Chickering for implementing the software used in our analysis of the Sewall and Shah (1968) data.

References

1. Aliferis C and Cooper G (1994) An evaluation of an algorithm for inductive learning of Bayesian belief networks using simulated data sets. In Proceedings of Tenth Conference on Uncertainty in Artificial Intelligence Seattle WA pages 8-14. Morgan Kaufmann
2. Becker S and LeCun Y (1989) Improving the convergence of backpropagation learning with second order methods In Proceedings of the 1988 Connectionist Models Summer School pages 29 -37. Morgan Kaufmann
3. Bernardo J and Smith A (1984) Bayesian Theory John Wiley and Sons New York
4. Buntine W (1991) Theory refinement on Bayesian networks In Proceedings of Seventh Conference on Uncertainty in Artificial Intelligence Los Angeles CA pages 52-60. Morgan Kaufmann
5. Buntine W (1994) Operations for learning with graphical models Journal of Artificial Intelligence Research 2:159-225
6. Cheeseman P and Stutz J (1995) Bayesian classification (AutoClass): Theory and results. In Fayyad U, Piatesky-Shapiro G, Smyth P and Uthurusamy R editors, Advances in Knowledge Discovery and Data Mining pages 153 180. AAAI Press Menlo Park CA
7. Chib S (1995) Marginal likelihood from the Gibbs output. Journal of the American Statistical Association, 90:1313-1321.
8. Chickering D (1996a) Learning Bayesian networks is NP complete. In Fisher D and Lenz H editors, Learning from Data, pages 121-130 SpringerVerlag.
9. Chickering D (1996b) Learning equivalence classes of Bayesian network structures. In Proceedings of Twelfth Conference on Uncertainty in Artificial Intelligence Portland OR Morgan Kaufmann
10. Chickering D and Heckerman D (1997) Efficient approximations for the marginal likelihood of Bayesian networks with hidden variables. Machine Learning,29: 181-212.
11. Cooper G (1995) Causal discovery from data in the presence of selection bias. In Proceedings of the Fifth International Workshop on Artificial Intelligence and
12. Statistics pages 140-150. Fort Lauderdale FL
13. Cooper G and Herskovits E (1992) A Bayesian method for the induction of probabilistic networks from data. Machine Learning 9:309-347.
14. Crawford S (1994) An application of the Laplace method to finite mixture distributions Journal of the American Statistical Association 89:259-267.

15. Dempster A Laird N and Rubin D (1977) Maximum likelihood from incomplete data via the EM algorithm. Journal of the Royal Statistical Society B 39:1-38.

16. DiCiccio T, Kass R, Raftery A and Wasserman L (July 1995) Computing Bayes factors by combining simulation and asymptotic approximations Technical Report 630, Department of Statistics, Carnegie Mellon University PA

17. Geiger D and Heckerman D (1994) Learning Gaussian networks. In Proceedings of Tenth Conference on Uncertainty in Artificial Intelligence Seattle WA pages 235-243. Morgan Kaufmann

18. Geiger D and Heckerman D (Revised February 1995). A characterization of the Dirichlet distribution applicable to learning Bayesian networks. Technical Report MSR-TR-94-16, Microsoft Research Redmond WA

19. Geiger D Heckerman D and Meek C (1996) Asymptotic model selection for directed networks with hidden variables. In Proceedings of Twelfth Conference on Uncertainty in Artificial Intelligence, Portland, OR. Pages 283-290. Morgan Kaufmann

20. Geman S and Geman D (1984) Stochastic relaxation Gibbs distributions and the Bayesian restoration of images. IEEE Transactions on Pattern Analysis and Machine Intelligence, 6:721-742.

21. Haughton D (1988) On the choice of a model to t data from an exponential family. Annals of Statistics 16:342-355.

22. Heckerman D (1995) A Bayesian approach for learning causal network. In Proceedings of Eleventh Conference on Uncertainty in Artificial Intelligence, Montreal QU pages 285-295. Morgan Kaufmann

23. Heckerman D and Geiger D (1996) (Revised November 1996) Likelihoods and priors for Bayesian networks. Technical Report MSR-TR-95-54 Microsoft Research, Redmond, WA

24. Heckerman D Geiger D and Chickering D (1995) Learning Bayesian networks: The combination of knowledge and statistical data. Machine Learning 20:197-243.

25. Herskovits E (1991) Computer-based probabilistic network construction. PhD thesis, Medical Information Sciences, Stanford University, Stanford CA

26. Jensen F, Lauritzen S and Olesen K (1990) Bayesian updating in recursive graphical models by local computations. Computational Statisticals Quarterly 4:269-282.

27. Kass R and Raftery A (1995) Bayes factors. Journal of the American Statistical Association, 90:773-795.

28. Kass R, Tierney L and Kadane J (1988) Asymptotics in Bayesian computation. In Bernardo J, DeGroot M, Lindley D and Smith A editors, Bayesian Statistics 3, pages 261-278, Oxford University Press

29. Kass R and Wasserman L (1995) A reference Bayesian test for nested hypotheses and its relationship to the Schwarz criterion. Journal of the American Statistical Association 90:928-934.

30. Madigan D, Garvin J and Raftery A (1995) Eliciting prior information to enhance the predictive performance of Bayesian graphical models. Communications in Statistics Theory and Methods 24:2271-2292.
31. Madigan D, Raftery A, Volinsky C and Hoeting J. (1996) Bayesian model averaging. In Proceedings of the AAAI Workshop on Integrating Multiple Learned Models, Portland, OR.
32. Madigan D and York J (1995) Bayesian graphical models for discrete data. International Statistical Review 63:215-232.
33. McLachlan G and Krishnan T (1997) The EM algorithm and extensions. Wiley
34. Meek C. (1995) Strong completeness and faithfulness in Bayesian networks. In Proceedings of Eleventh Conference on Uncertainty in Artificial Intelligence. Montreal QU pages 411-418. Morgan Kaufmann
35. Meng X and Rubin D (1991) Using EM to obtain asymptotic variance-covariance matrices: The SEM algorithm. Journal of the American Statistical Association 86:899-909.
36. Neal R (1993) Probabilistic inference using Markov chain Monte Carlo methods. Technical Report CRG-TR-93-1. Department of Computer Science, University of Toronto
37. Raftery A (1995) Bayesian model selection in social research. In Marsden P, editor, Sociological Methodology. Blackwells, Cambridge MA.
38. Raftery A (1996) Hypothesis testing and model selection chapter 10. Chapman and Hall
39. Rissanen J (1987) Stochastic complexity (with discussion). Journal of the Royal Statistical Society Series B 49:223-239 and 253-265.
40. Robins J (1986) A new approach to causal inference in mortality studies with sustained exposure results. Mathematical Modelling 7:1393-1512.
41. Rubin D (1978) Bayesian inference for causal effects: The role of randomization. Annals of Statistics 6:34-58.
42. Russell S, Binder J, Koller D and Kanazawa K. (1995) Local learning in probabilistic networks with hidden variables. In Proceedings of the Fourteenth International Joint Conference on Artificial Intelligence Montreal QU pages 1146-1152. Morgan Kaufmann San Mateo CA
43. Scheines R, Spirtes P, Glymour C and Meek C (1994)Tetrad II Users Manual. Lawrence Erlbaum, Hillsdale NJ
44. Schwarz G (1978) Estimating the dimension of a model. Annals of Statistics, 6:461-464.
45. Sewell W and Shah V (1968) Social class parental encouragement and educational aspirations. American Journal of Sociology 73:559-572.
46. Singh M and Valtorta M (1993) An algorithm for the construction of Bayesian network structures from data. In Proceedings of Ninth Conference on Uncertainty in Artificial Intelligence, Washington DC pages 259-265. Morgan Kaufmann
47. Spiegelhalter D and Lauritzen S (1990) Sequential updating of conditional probabilities on directed graphical structures. Networks 20:579-605.

48. Spirtes P, Glymour C and Scheines R (1993) Causation, Prediction and Search. SpringerVerlag New York
49. Spirtes P and Meek C (1995) Learning Bayesian networks with discrete variables from data In Proceedings of First International Conference on Knowledge Discovery and Data Mining Montreal QU Morgan Kaufmann
50. Spirtes P, Meek C and Richardson T (1995) Causal inference in the presence of latent variables and selection bias. In Proceedings of Eleventh Conference on Uncertainty in Artificial Intelligence Montreal QU pages 499-506. Morgan Kaufmann
51. Thiesson B (1997) Score and information for recursive exponential models with incomplete data. Technical report Institute of Electronic System, Aalborg University, Aalborg Denmark
52. Verma T and Pearl J (1990) Equivalence and synthesis of causal models. In Proceedings of Sixth Conference on Uncertainty in Artificial Intelligence Boston MA pages 220-227. Morgan Kaufmann
53. Winkler R (1967)The assessment of prior distributions in Bayesian analysis. American Statistical Association Journal 62:776-800.

2 A Tutorial on Learning Causal Influence

Richard E. Neapolitan, Xia Jiang
Northeastern Illinois University
RE-Neapolitan@neiu.edu, xjiang@cbmi.pitt.edu

Abstract

In the 1990's related research in artificial intelligence, cognitive science, and philosophy resulted in a method for learning causal relationships from passive data when we have data on at least four variables. We illustrate the method using a few simple examples. Then we present recent research showing that we can even learn something about causal relationships when we have data on only two variables.

2.1 Introduction

Christensen [1] says "Causality is not something that can be established by data analysis. Establishing causality requires logical arguments that go beyond the realm of numerical manipulation." The argument is that if, for example, we determine that smoking and lung cancer are correlated, we cannot necessarily deduce that smoking causes lung cancer. The following are other causal explanations for smoking and lung cancer being correlated: lung cancer could cause smoking; lung cancer and smoking could cause each other via a causal feedback loop; lung cancer and smoking could have a hidden common cause such as a genetic defect; and lung cancer and smoking could cause some other condition, and we have sampled from a population in which all individuals have this condition (This is called selection bias.). All five causal explanations are shown in Figure 2.1.

This argument that causal relationships cannot be learned from passive data is compelling when we have data on only two variables (e.g. smoking and lung cancer). However, in the 1990's related research in artificial intelligence, cognitive science, and philosophy [6, 11, 13, 17] resulted in a method for learning causal relationships when we have data on at least four variables (or three variables if we have

R. E. Neapolitan and X. Jiang: *A Tutorial on Learning Causal Influence*, StudFuzz **194**, 29–71 (2006)
www.springerlink.com

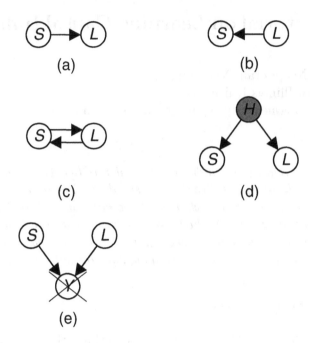

Figure 2.1: All five causal relationships could account for smoking (S) and lung cancer (L) being correlated.

a time ordering of the events). This method is discussed in detail in [11] and [17]. In Section 2 we illustrate the method using a few simple examples. Then in Section 3 we present recent research showing that we can even learn something about causal relationships when we have data on only two variables. In the current section we present some necessary preliminary concepts including a definition of causal networks.

2.1.1   Causation

A common way to learn (perhaps define) causation is via manipulation experiments. We say we **manipulate** X when we force X to take some value, and we say X **causes** Y if there is some manipulation of X that leads to a change in the probability distribution of Y. A manipulation consists of a randomized controlled experiment (**RCE**) using some specific population of entities (e.g. individuals with chest pain) in some specific context (E.g., they currently receive no chest pain medication and they live in a particular geographical area.). The causal relationship discovered is then relative to this population and this context.

Let's discuss how the manipulation proceeds. We first identify the population of entities we wish to consider. Our variables are features of these entities. Next we ascertain the causal relationship we wish to investigate. Suppose we are trying to determine if variable X is a cause of variable Y. We then sample a number of entities from the population. For every entity selected, we manipulate the value of X so that each of its possible values is given to the same number of entities (If X is continuous, we choose the values of X according to a uniform distribution.). After the value of X is set for a given entity, we measure the value of Y for that entity. The more the resultant data shows a dependency between X and Y the more the data supports that X causally influences Y. The manipulation of X can be represented by a variable M that is external to the system being studied. There is one value mi of M for each value xi of X, the probabilities of all values of M are the same, and when M equals mi, X equals xi. That is, the relationship between M and X is deterministic. The data supports that X causally influences Y to the extent that the data indicates $P(yi|mj) \neq P(yi|mk)$ for $j \neq k$. Manipulation is actually a special kind of causal relationship that we assume exists primordially and is within our control so that we can define and discover other causal relationships. Figure 2.2 depicts a manipulation experiment in which we are trying to determine whether smoking causes lung cancer. In that figure, $P(m1) = .5$ means the probability of being selected for smoking is .5, and$P(s1|m1) = 1$ means the probability of smoking ($s1$) is 1 if the person is selected for smoking. Similarly, the $P(s2|m2) = 1$ means the probability of not smoking ($s2$) is 1 if the person is selected for not smoking.

We do not really want to manipulate people and make them smoke. We will see in the next section that it is possible to learn something about whether smoking causes lung cancer from passive data alone.

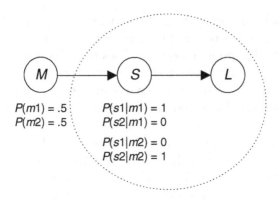

$P(m1) = .5$ $P(s1|m1) = 1$
$P(m2) = .5$ $P(s2|m1) = 0$

$P(s1|m2) = 0$
$P(s2|m2) = 1$

Figure 2.2: A manipulation experiment to determine whether smoking (S) causes lung cancer (L).

First we need to define causal networks.

2.1.2   Causal networks

If we create a causal DAG (directed acyclic graph) $\mathbb{G}$ and assume the observed probability distribution P of the variables/nodes in the DAG satisfies the Markov condition with $\mathbb{G}$, we say we say we are making the **causal Markov assumption**, and $(\mathbb{G}, P)$ is called a **causal network** [10]. A **causal DAG** is a DAG in which there is an edge from X to Y if and only if X is a direct cause of Y. By a 'direct cause' we mean a manipulation of X results in a change in the probability distribution of Y, and there is no subset of variables W of the set of variables in the DAG such that if we knew the values of the variables in W, a manipulation of X would no longer change the probability distribution of Y. A probability distribution P satisfies the **Markov condition** with a DAG $\mathbb{G}$ if the probability of each variable/node in the DAG is independent of its nondescendents conditional on its parents. We will use the notation $I(X, Y|Z)$ to denote that X is independent of Y conditional on Z. That is, $I(X, Y|Z)$ holds if for all values x, y, and

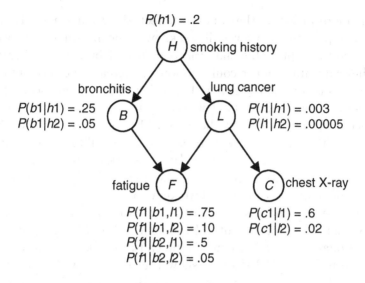

$P(h1) = .2$

H smoking history

bronchitis

$P(b1|h1) = .25$
$P(b1|h2) = .05$

B

lung cancer

L

$P(l1|h1) = .003$
$P(l1|h2) = .00005$

fatigue F

C chest X-ray

$P(f1|b1,l1) = .75$
$P(f1|b1,l2) = .10$
$P(f1|b2,l1) = .5$
$P(f1|b2,l2) = .05$

$P(c1|l1) = .6$
$P(c1|l2) = .02$

Figure 2.3: A Causal Network

z of X, Y, and Z we have

$$P(x|y, z) = P(x|z).$$

Similarly, if A is a set of variables and X is independent of the variables in A conditional on Z, we write $I(X, \mathsf{A}|Z)$.

Consider the causal network in Figure 2.3. The causal Markov assumption for that network entails the following conditional independencies:

$$I(B, \{L, C\}|H) \qquad I(F, \{H, C\}|\{L, B\})$$
$$I(L, B|H) \qquad I(C, \{H, B, F\}|L).$$

Why should we make the causal Markov assumption? A study in [7] supplies experimental evidence for it. We offer the following brief intuitive justification. Given the causal relationship in Figure 2.3, we would not expect bronchitis and lung cancer to be independent because if someone had lung cancer it would make it more probable that they smoked (since smoking is a cause of lung cancer), which would make it more probable that another effect of smoking, namely bronchitis, was present. However, if we knew someone smoked, it would already be more probable that the person had bronchitis. Learning that they had lung cancer could no longer increase the probability of smoking (This probability is now 1.), which means it can't change the probability of bronchitis. That is, the variable H shields B from the influence of L, which is what the causal Markov condition says. Similarly, a positive chest X-ray increases the probability of lung cancer, which in turn increases the probability of smoking, which in turn increases the probability of bronchitis. So a chest X-ray and bronchitis are not independent. However, if we knew the person had lung cancer, the chest X-ray could not change the probability of lung cancer and thereby change the probability of bronchitis. So B is independent of C conditional on L, which is what the causal Markov condition says.

Notice in Figure 2.3 that we show the probability distribution of each variable conditional on the values of its parents. It is a theorem that if P satisfies the Markov condition with $\mathbb{G}$, then P is equal to the product of its conditional distributions in $\mathbb{G}$ (See [10,11] for the proof.). So in a causal network we specify the probability distribution by showing the conditional distributions. If the network is sparse, this is far more efficient than listing every value in the probability distribution. For example, if there are 100 binary variables, there are 2^{100} values in the joint probability distribution. However, if each variable has at most three parents, less than 800 values determine the conditional distributions.

The following conditions must be satisfied for the causal Markov condition to hold:

1. If X causes Y, we must draw an edge from X to Y unless all causal paths from X to Y are mediated by observed variables.

2. There must be no causal feedback loops.

3. There must be no hidden common causes.

4. Selection bias must not be present.

These conditions are discussed in detail in [11]. Perhaps the condition that is most violated is that there can be no hidden common

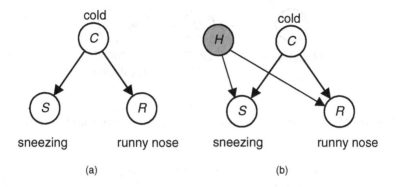

Figure 2.4: The causal Markov assumption would not hold for the DAG in (a) if there is a hidden common cause as depicted in (b).

causes. We discuss that condition further here. Suppose we draw a causal DAG containing the variables cold (C), sneezing (S), and runny nose (R). Then since a cold can cause both sneezing and a runny nose and neither of these conditions can cause each other, we would draw the DAG in Figure 2.4 (a). The causal Markov condition for that DAG would entail $I(S, R|C)$. However, if there were a hidden common cause of S and R as depicted in (b), this conditional independency would not hold because even if the value of C were know, S would change the probability of H, which in turn would change the probability of R. Indeed, there is another cause of sneezing and runny nose, namely hay fever. So when making the causal Markov assumption, we must be certain that we have identified all common causes.

Consider now the causal DAG in Figure 2.5. The pharmaceutical company Merck developed the drug finasteride which lowers dehydro-testosterone (DHT) levels in men, and DHT is believed to be the hormone responsible for baldness in men. However, DHT is also necessary for erectile function, as shown by a study in [7]. Merck wanted to market finasteride as a hair regrowth treatment, but they feared the side effect of erectile dysfunction. So Merck conducted a large scale manipulation study [8], and found that finasteride has a causal effect on DHT level but has no causal effect on erectile dysfunction.

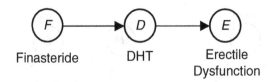

Figure 2.5: Finasteride and erectile function are independent.

How could this be when the causal relationships among finasteride (F), DHT (D), and erectile dysfunction (E) have clearly been found to be those depicted in Figure 2.5? We would expect a causal mediary to transmit an effect from its antecedent to its consequence, but in this case it does not. The explanation is that finasteride cannot lower DHT levels beyond a certain threshold level, and that level is all that is needed for erectile function. So we have $I(F, E)$.

The Markov condition does not entail $I(F, E)$ for the causal DAG in Figure 2.5. It only entails $I(F, E|D)$. When a probability distribution has a conditional independency that is not entailed by the Markov condition, the faithfulness assumption does not hold. If we create a causal DAG $\mathbb{G}$ and assume

1. $(\mathbb{G}, P)$ satisfies the Markov condition,

2. All conditional independencies in the observed distribution P are entailed by the Markov condition in $\mathbb{G}$,

then we say we are making the **causal faithfulness assumption**.

Besides those conditions already stated for the causal Markov assumption to hold, the following additional conditions must be satisfied for the causal faithfulness assumption to hold:

1. We can't have 'unusual' causal relationships as in the finasteride example.

2. We cannot draw an edge from X to Y if for every causal path from X to Y there is a causal mediary in the set of observed variables.

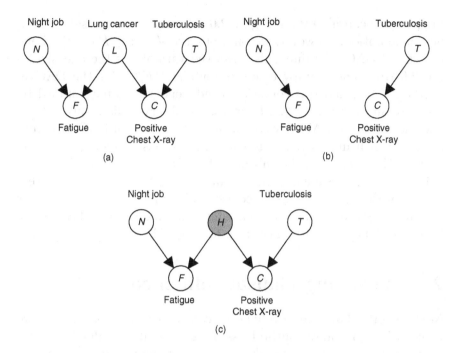

Figure 2.6: If the causal relationships are those shown in (a), P is not faithful to the DAG in (b), but P is embedded faithfully in the DAG in (c).

It seems the main exception to the causal faithfulness assumption (and the causal Markov assumption) is the presence of hidden common causes. Our next assumption eliminates that exception. If we assume the observed probability distribution P of the variables is embedded faithfully in a causal DAG containing the variables, we say we are making the **causal embedded faithfulness assumption**. Suppose we have a probability distribution P of the variables in a set V, V is a subset of W, and $\mathbb{G}$ is a DAG whose set of nodes is W. Then P is **embedded faithfully** in W if all and only the conditional independencies in P are entailed by the Markov condition applied to W and restricted to the nodes in V.

Next we illustrate the causal embedded faithfulness assumption.

Suppose the causal DAG in Figure 2.6 (a) satisfies the causal faithfulness assumption. However, we only observe $V = \{N, F, C, T\}$. Then the causal DAG containing the observed variables is the one in Figure 2.6 (b). The Markov condition entails $I(F, C)$ for the DAG in Figure 2.6 (b), and this conditional independency is not entailed by the DAG in Figure 2.6 (a). Therefore, the observed distribution $P(V)$ does not satisfy the Markov condition with the causal DAG in Figure 2.6 (b), which means the causal faithfulness assumption is not warranted. However, $P(V)$ is embedded faithfully in the DAG in Figure 2.6 (c). So the causal embedded faithfulness assumption is warranted. Note that this example illustrated a situation in which we identify four variables and two of them have a hidden common cause. That is, we have not identified lung cancer as a feature of humans.

2.2   Learning Causal Influences

Next we show how causal influences can be learned from data if we make either the causal faithfulness or the causal embedded faithfulness assumption. We assume that we have a random sample of entities and know the values of the variables of interest for the entities in the sample. From this sample, we have deduced the conditional independencies among the variables. A method for doing this is described in [11] and [17]. Our confidence in the causal influences we conclude is no greater than our confidence in these conditional independencies.

2.2.1   Making the Causal Faithfulness Assumption

We assume here that the variables satisfy the causal faithfulness assumption and we know the conditional independencies among the variables. Given these assumptions, we present a sequence of examples showing how causal influences can be learned.

Example 1 *Suppose* V *is our set of observed variables,* $V = \{X, Y\}$, *and our set of conditional independencies is*

$$\{I(X, Y)\}.$$

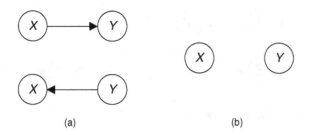

(a)   (b)

Figure 2.7: If the set of conditional independencies is $\{I(X,Y)\}$, we must have the causal DAG in (b), whereas if it is $\varnothing$, we must have one of the causal DAGs in (a).

Then we cannot have either of the causal DAGs in Figure 2.7 (a). The reason is that the Markov condition, applied to these DAGs, does not entail that X and Y are independent, which means the causal faithfulness assumption is not satisfied. So we must have the causal DAG in Figure 2.7 (b). We conclude X and Y have no causal influence on each other.

Example 2 *Suppose $V = \{X,Y\}$ and our set of conditional independencies is the empty set*

$$\varnothing.$$

Then we cannot have the causal DAG in Figure 2.7 (b). The reason is that the Markov condition, applied to this DAG, entails that X and Y are independent, which means the causal Markov assumption would not be satisfied. So we must have one of the causal DAGs in Figure 2.7 (a). We conclude either X causes Y or Y causes X.

Example 3 *Suppose $V = \{X,Y,Z\}$ and our set of conditional independencies is*

$$\{I(X,Y)\}.$$

Then there can be no edge between X and Y in the causal DAG owing to the reason given in Example 1. Furthermore, there must be edges between X and Z and between Y and Z owing to the reason given in Example 2. We cannot have any of the causal DAGs in Figure

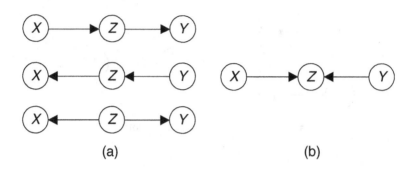

Figure 2.8: If the set of conditional independencies is $\{I(X,Y)\}$, we must have the causal DAG in (b).

2.8 (a).   The reason is that the Markov condition, applied to these DAGs, entails $I(X,Y|Z)$, and this conditional independency is not present. So the Markov condition would not be satisfied. Furthermore, the Markov condition, applied to these DAGs, does not entail $I(X,Y)$. So the causal DAG must be the DAG in Figure 2.8 (b). We conclude that X and Y each cause Z.

Example 4 *Suppose* $\mathsf{V} = \{X,Y,Z\}$ *and our set of conditional independencies is*

$$\{I(X,Y|Z)\}.$$

Then owing to reasons similar to those given before, the only edges in the causal DAG must be between X and Z and between Y and Z. We cannot have the causal DAG in Figure 2.8 (b). The reason is that the Markov condition, applied to this DAG, entails $I(X,Y)$, and this conditional independency is not present. So the Markov condition would not be satisfied. So we must have one of the causal DAGs in Figure 2.8 (a).

We now state the following theorem, whose proof can be found in [11]. At this point your intuition should suspect that it is true.

Theorem 1 *If $(\mathbb{G}, P)$ satisfies the faithfulness condition, then there is an edge between X and Y if and only if X and Y are not conditionally independent given any set of variables.*

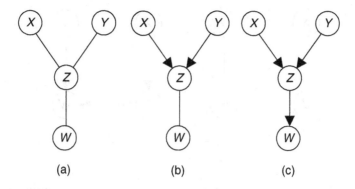

(a)   (b)   (c)

Figure 2.9:   If the set of conditional independencies is $\{I(X,Y),\quad I(W,\{X,Y\}|Z)\}$, we must have the causal DAG in (c).

Example 5 *Suppose* $\mathsf{V} = \{X,Y,Z,W\}$ *and our set of conditional independencies is*

$$\{I(X,Y),\quad I(W,\{X,Y\}|Z)\}.$$

Owing to Theorem 1, the links (edges without regard for direction) must be as shown in Figure 2.9 (a). We must have the directed edges shown in Figure 2.9 (b) because we have $I(X,Y)$. Therefore, we must also have the directed edge shown in Figure 2.9 (b) because we do not have $I(W,X)$. We conclude X and Y each cause Z and Z causes W.

Example 6 *Suppose* $\mathsf{V} = \{X,Y,Z,W\}$ *and our set of conditional independencies is*

$$\{I(X,\{Y,W\}),\quad I(Y,\{X,Z\})\}.$$

Owing to Theorem 1, we must have the links shown in Figure 2.10 (a). Now if we have the chain $X \to Z \to W$, $X \leftarrow Z \leftarrow W$, or $X \leftarrow Z \to W$, then we would not have $I(X,W\}$, which is an independency that can be readily deduced from $I(X,\{Y,W\})$. So we must have the chain $X \to Z \leftarrow W$. Similarly, we must have chain $Y \to W \leftarrow Z$. So our causal graph must be the one in Figure 2.10 (b). However,

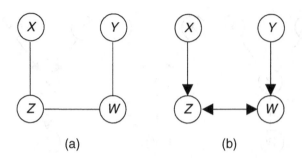

(a)   (b)

Figure 2.10: If the set of conditional independencies is $\{I(X,\{Y,W\}),\quad I(Y,\{X,Z\})\}$, we must have the causal graph in (b).

this graph is not a DAG. The problem here is that this probability distribution does not admit a faithful DAG representation, which tells us we cannot make the causal faithfulness assumption. In the next subsection we will revisit this example while making only the causal embedded faithfulness assumption.

2.2.2 Assuming Only Causal Embedded Faithfulness

Previously, we mentioned that the most problematic assumption in the causal faithfulness assumption is that there must be no hidden common causes, and we eliminated that problem with the causal embedded faithfulness assumption. Let's see how much we can learn when making only this assumption.

Example 7 *In Example 3 we had* $V = \{X, Y, Z\}$ *and the set of conditional independencies*

$$\{I(X,Y)\}.$$

Under the causal faithfulness assumption, we concluded that X and Y each cause Z. However, the probability distribution is embedded faithfully in all the DAG in Figure 2.11. So it could be that X causes Z or it could be that X and Z have a hidden common cause. The same holds for Y and Z.

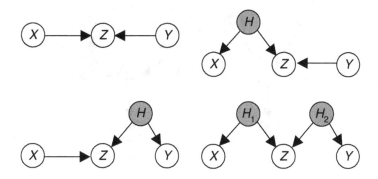

Figure 2.11: If we make the causal embedded faithfulness assumption and our set of conditional independencies is $\{I(X,Y)\}$, the causal relationships could be the ones in any of these DAGs.

While making only the more reasonable causal embedded faithfulness assumption, we were not able to learn any causal influences in the previous example. Can we ever learn a causal influence while making this assumption? The next example shows that we can.

Example 8 *In Example 5 we had* $\mathsf{V} = \{X, Y, Z, W\}$ *and the set of conditional independencies*

$$\{I(X,Y),\quad I(W,\{X,Y\}|Z)\}.$$

The probability distribution P is embedded faithfully in the DAGs in Figure 2.12 (a) and (b). However, it is not embedded faithfully in the DAGs in Figure 2.12 (c) or (d). The reason is that these latter DAGs entail $I(X,W)$, *and we do not have this conditional independency. That is, the Markov condition says X must be independent of its non-descendents conditional on its parents. Since X has no parents, this means X must simply be independent of its nondescendents, and W is one of its nondescendents. We conclude that Z causes W.*

Example 9 *In Example 6 we had* $\mathsf{V} = \{X, Y, Z, W\}$ *and the set of conditional independencies*

$$\{I(X,\{Y,W\}),\quad I(Y,\{X,Z\})\}.$$

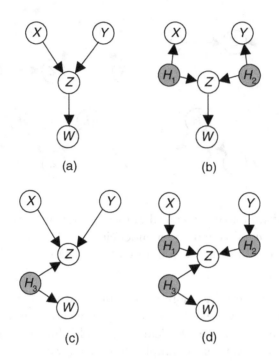

(a)   (b)

(c)   (d)

Figure 2.12:   If our set of conditional independencies is $\{I(X,Y),\quad I(W,\{X,Y\}|Z)\}$, then P is embedded faithfully in the DAGs in (a) and (b) but not in the DAGs in (c) and (d).

Recall that we obtained the graph in Figure 2.13 (a) when we tried to find a DAG faithful to the probability distribution P. We concluded that P does not admit a faithful DAG representation. On the other hand, P is embedded faithfully in the DAGs in Figure 2.13 (b) and (c). We conclude that Z and W have a hidden common cause.

Example 10 *Suppose we have these variables:*

R: Parent's smoking history

A: Alcohol consumption

S: Smoking behavior

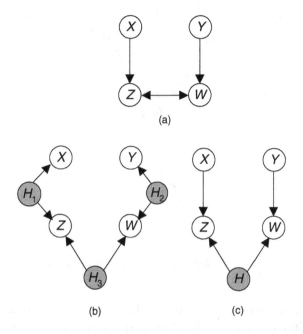

Figure 2.13:   If our set of conditional independencies is $\{I(X, \{Y, W\}),\quad I(Y, \{X, Z\})$, we can conclude that Z and W have a hidden common cause.

L: *Lung Cancer*

Suppose further we learn the following conditional independencies from data:

$$\{I(R, A),\quad I(L, \{R, A\}|S)\}.$$

We conclude the causal relationships in Figure 2.14. In that figure the edges $R \rightarrow S$ and $A \rightarrow S$ mean that the first variable causes the second, or the two variables have a hidden common cause, or the first causes the second and they have a hidden common cause. The edge $S \rightarrowtail L$ means S causes L and they do not have a hidden common cause. We conclude smoking causes lung cancer. This example is only for the sake of illustration. We know of no data set indicating these conditional independencies.

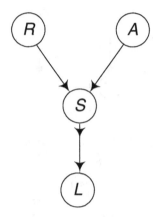

Figure 2.14: S has a causal influence on L.

Based on considerations such as those illustrated in the previous examples, Spirtes et al. [17] developed an algorithm that finds the causal DAG faithful to P from the conditional independencies in P when the causal faithfulness assumption is made. Meek [9] proved the correctness of the algorithm. Spirtes et al. [17] further developed an algorithm that learns causal influences from the conditional independencies in P when the causal embedded faithfulness assumption is made. They conjecture that it finds all possible causal influences among the observed variables. The algorithm is also described in detail in [11].

Next we show two studies that use the method just described to learn causal influences.

Example 11 *Using the data base collected by the U.S. News and World Record magazine for the purpose of college ranking, Druzdzel and Glymour [4] analyzed the influences that affect university student retention rate. By 'student retention rate' we mean the percent of entering freshmen who end up graduating from the university at which they initially matriculate. Low student retention rate is a major concern at many American universities as the mean retention rate over all American universities is only 55%.*

The data base provided by the U.S. News and World Record magazine contains records for 204 United States universities and colleges identified as major research institutions. Each record consists of over 100 variables. The data was collected separately for the years 1992 and 1993. Druzdzel and Glymour [4] selected the following eight variables as being most relevant to their study:

Variable	What the Variable Represents
grad	Fraction of entering students graduating from the institution
rejr	Fraction of applicants who are not offered admission
tstsc	Average standardized score of incoming students
tp10	Fraction of incoming students in the top 10% in high school
acpt	Fraction of students accepting the institution's admission offer
spnd	Average educational and general expenses per student
sfrat	Student/faculty ratio
salar	Average faculty salary

Druzdzel and Glymour [4] used Tetrad II [16] to learn causal influences from the data. Tetrad II uses the previously mentioned algorithm developed by Spirtes et al. [17] to learn causal structure from data. Tetrad II allows the user to enter a significance level. A significance level of α means the probability of rejecting a conditional independency hypothesis, when it it is true, is α. Therefore, the smaller the value α, the less likely we are to reject a conditional independency, and therefore the sparser our resultant graph. Figure 2.15 shows the graphs, which Druzdzel and Glymour [4] learned from U.S. News and World Record's 1992 data base using significance levels of .2, .1, .05, and .01. In those graphs, an edge $X \rightarrow Y$ indicates either X has a causal influence on Y, or X and Y have a hidden common cause or both; an edge $X \leftrightarrow Y$ indicates X and Y have a hidden common cause; and an edge $X \rightarrowtail Y$ indicates X has a causal influence on Y and they do not have a hidden common cause.

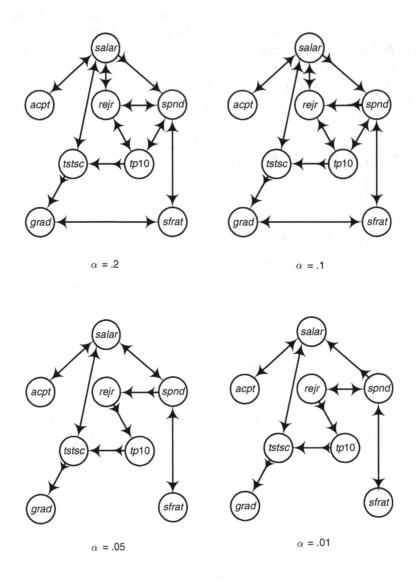

Figure 2.15: The graphs Tetrad II learned from *U.S. News and World Record*'s 1992 data base.

Although different graphs were obtained at different levels of signifi-cance, all the graphs in Figure 2.15 show that average standardized test score (tstsc) has a direct causal influence on graduation rate (grad), and no other variable has a direct causal influence on grad. The results for the 1993 data base were not as overwhelming, but they too indicated tstsc to be the only direct causal influence of grad.

To test whether the causal structure may be different for top research universities, Druzdzel and Glymour [4] repeated the study using only the top 50 universities according to the ranking of U.S. News and World Report. The results were similar to those for the complete data base.

These result indicate that, although factors such as spending per student and faculty salary may have an influence on graduation rates, they do this only indirectly by affecting the standardized test scores of matriculating students. If the results correctly model reality, retention rates can be improved by bringing in students with higher test scores in any way whatsoever. Indeed, in 1994 Carnegie Mellon changed its financial aid policies to assign a portion of its scholarship fund on the basis of academic merit. Druzdzel and Glymour [4] note that this resulted in an increase in the average test scores of matriculating freshman classes and an increase in freshman retention.

Before closing, we note that the notion that average test score has a causal influence on graduation rate does not fit into common notions of causation such as the one concerning manipulation. For example, if we manipulated a university's average test score by accessing the testing agency's database and changing the scores of the university's students to much higher values, we would not expect the university's graduation rate to increase. Rather this study indicates that test score is a near perfect indicator of some other variable, which we can call 'graduation potential'.

The next example, taken from [12] illustrates problems one can encounter when inferring causation from passive data.

Example 12 *Scarville et al. [15] provide a data base obtained from a survey in 1996 of experiences of racial harassment and discrimination of military personnel in the United States Armed Forces. Surveys were* *distributed to 73,496 members of the U.S. Army, Navy, Marine Corps, Air Force and Coast Guard. The survey sample was selected using a nonproportional stratified random sample in order to ensure adequate representation of all subgroups. Usable surveys were received from 39,855 service members (54%). The survey consisted of 81 questions related to experiences of racial harassment and discrimination and job attitudes. Respondents were asked to report incidents that had occurred during the previous 12 months. The questionnaire asked participants to indicate the occurrence of 57 different types of racial/ethnic harassment or discrimination. Incidents ranged from telling offensive jokes to physical violence, and included harassment by military personnel as well as the surrounding community. Harassment experienced by family members was also included.*

Neapolitan and Morris [12] used Tetrad III in an attempt to learn causal influences from the data base. For their analysis, 9640 records (13%) were selected which had no missing data on the variables of interest. The analysis was initially based on eight variables. Similar to the situation concerning university retention rates, they found one causal relationship to be present regardless of the significance level. That is, they found that whether the individual held the military responsible for the racial incident had a direct causal influence on the individual's race. Since this result made no sense, they investigated which variables were involved in Tetrad III learning this causal influence. The five variables involved are the following:

Variable	What the Variable Represents
race	*Respondent's race/ethnicity*
yos	*Respondent's years of military service*
inc	*Did respondent experience a racial incident?*
rept	*Was incident reported to military personnel?*
resp	*Did respondent hold military responsible for incident?*

The variable race consisted of five categories: White, Black, Hispanic, Asian or Pacific Islander, and Native American or Alaskan Native.

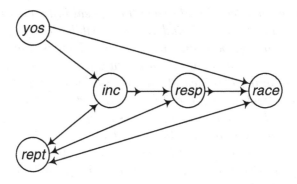

Figure 2.16: The graph Tetrad III learned from the racial harassment survey at the .01 significance level.

Respondents who reported Hispanic ethnicity were classified as Hispanic, regardless of race. Respondents were classified based on self-identification at the time of the survey. Missing data were replaced with data from administrative records. The variable yos was classified into four categories: 6 years or less, 7-11 years, 12-19 years, and 20 years or more. The variable inc was coded dichotomously to indicate whether any type of harassment was reported on the survey. The variable rept indicates responses to a single question concerning whether the incident was reported to military and/or civilian authorities. This variable was coded 1 if an incident had been reported to military officials. It was coded 0 if an individual experienced no incident, did not report the incident, or only reported the incident to civilian officials. The variable resp indicates responses to a single question concerning whether the respondent believed the military to be responsible for an incident of harassment. This variable was coded 1 if the respondent indicated that the military was responsible for some or all of a reported incident. If the respondent indicated no incident, unknown responsibility, or that the military was not responsible, the variable was coded 0.

They reran the experiment using only these five variables, and again at all levels of significance, they found that resp had a direct causal influence on race. In all cases, this causal influence was learned because rept and yos were found to be probabilistically independent, and there was no edge between race and inc. That is, the causal connection between race and inc is mediated by other variables. Figure 2.16 shows the graph obtained at the .01 significance level. The edges yos → inc and rept → inc are directed towards inc because yos and rept were found to be independent. The edge yos → inc resulted in the edge inc ⟼ resp being directed the way it was, which in turn resulted in resp ⟼ race being directed the way it was. If there had been an edge between inc and race, the edge between resp and race would not have been directed.

It seems suspicious that no direct causal connection between race and inc was found. Recall, however, that these are the probabilistic relationships among the responses; they are not necessarily the probabilistic relationships among the actual events. There is a problem with using responses on surveys to represent occurrences in nature because subjects may not respond accurately. This is called response bias. Let's assume race is recorded accurately. The actual causal relationship between race, inc, and says_inc may be as shown in Figure 2.17. By inc we now mean whether there really was an incident, and by says_inc we mean the survey response. It could be that races, which experienced higher rates of harassment, were less likely to report the incident, and the causal influence of race on says_inc through inc was negated by the direct influence of race on inc. The previous conjecture is substantiated by another study. Stangor et al. [18] examined the willingness of people to attribute a negative outcome to discrimination when there was evidence that the outcome might be influenced by bias. They found that minority members were more likely to attribute the outcome to discrimination when responses were recorded privately, but less likely to report discrimination when they had to express their opinion publicly and there was a member of the non-minority group present. This suggests that while minorities are more likely to perceive the situation as due to discrimination, they are less likely to report it publicly. Although the survey of military personnel was intended to be confidential, minority members in the military may have felt uncomfortable reporting incidents of discrimination.

Tetrad III allows the user to enter a temporal ordering. So Neapoli-

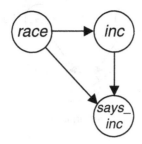

Figure 2.17: Possible causal relationships among race, incidence of harassment, and saying there is an incident of harassment.

tan and Morris [12] could have put race first in such an ordering to avoid it being an effect of another variable. However, one should do this with caution. The fact that the data strongly supports that race is an effect indicates there is something wrong with the data, which means one should be dubious of drawing any conclusions from the data. In the present example, Tetrad III actually informed them that they could not draw causal conclusions from the data when they made race a root. That is, when they made race a root, Tetrad III concluded there is no consistent orientation of the edge between race and resp.

2.2.3   Assuming Causal Embedded Faithfulness with Selection Bias

In the previous subsection we deduced causal influences assuming selection bias is not present. Here we relax that assumption. If we assume the probability distribution P of the observed variables is embedded faithfully in a causal DAG containing the variables, but that possibly selection bias is present when we sample, we say we are making the **causal embedded faithfulness assumption with selection bias**.

Before showing an example of causal learning under this assumption, let's discuss selection bias further. Recall in Section 2.1.2 we mentioned that the pharmaceutical company Merck noticed that its

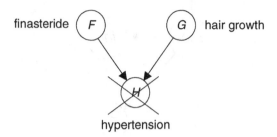

finasteride (F)   (G) hair growth

H

hypertension

Figure 2.18: The instantiation of H creates a dependency between F and G.

drug finasteride appeared to cause hair regrowth. Now suppose finasteride (F) and apprehension about lack of hair regrowth (G) are both causes of hypertension (H), and Merck happened to be observing individuals who had hypertension. We say a node is **instantiated** when we know its value for the entity currently being modeled. So we are saying H is instantiated to the same value for all entities in the population we are observing. This situation is depicted in Figure 2.18, where the cross through H means the variable is instantiated. Ordinarily, the instantiation of a common effect creates a dependency between its causes because each cause explains away the occurrence of the effect, thereby making the other cause less likely. Psychologists call this **discounting**. So, if this were the case, discounting would explain the correlation between F and G. As mentioned previously, this type of dependency is called **selection bias**. This example is only for the sake of illustration. There is no evidence that finasteride causes hypertension, and it apparently does cause hair regrowth by lowering DHT levels.

The next example shows that we can learn causal influences even when selection bias may be present.

Example 13 *Suppose we have the same variables and distribution as in Examples 5 and 8, and now we assume selection bias may be present. Recall* $\mathsf{V} = \{X, Y, Z, W\}$ *and our set of conditional independencies in the observed probability distribution P (the one obtained*

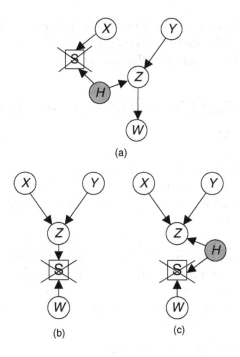

(a)

(b)   (c)

Figure 2.19:   If the observed set of conditional independencies is $\{I(X,Y),\quad I(W,\{X,Y\}|Z)\}$, the actual probability distribution could only be embedded faithfully in the DAG in (a).

when selection bias may be present) was

$$\{I(X,Y),\quad I(W,\{X,Y\}|Z)\}.$$

In Example 8, when making the assumption of causal embedded faithfulness, we concluded that Z causes W. Next we show that we can conclude this even when we only make the causal embedded faithfulness assumption with selection bias. The actual probability distribution P' (the one obtained when S is not instantiated and so there is no selection bias) could be embedded faithfully in the causal DAG in Figure 2.19 (a), but it could not be embedded faithfully in the casual DAG in Figure 2.19 (b) or (c). The reason is that, owing to the instantiation of S, the DAG in Figure 2.19 (b) does not entail $I(X,Y)$ in *the observed distribution P, and the DAG in Figure 2.19 (c) does not entail $I(W, \{X, Y\}|Z)$ in the observed distribution P. So we can still conclude that Z has a causal influence on W.*

2.3   Learning Causation From Data on Two Variables

The previous section discussed learning causal structure from the conditional independencies in the joint probability distribution of the variables. These conditional independencies are obtained from the data in a random sample. Another way to learn causal structure is to develop a scoring function *score* (called a **scoring criterion**) that assigns a value *score*(data, $\mathbb{G}$) to each causal DAG under consideration based directly on the data. Ideally we want a scoring criterion to be consistent. First we discuss consistency and some other preliminary concepts. Then we apply the method to causal learning.

2.3.1   Preliminary Concepts

A **DAG model** consists of a DAG $\mathbb{G} = (V, E)$, where V is a set of random variables, and a parameter set F whose members determine conditional probability distributions for the DAG, such that for every permissible assignment of values to the members of F, a joint probability distribution of V is given by the product of these conditional distributions and this joint probability distribution satisfies the Markov condition with the DAG. For simplicity, we ordinarily denote a DAG model using only $\mathbb{G}$ (i.e. we do not show F.) Probability distribution P is **included** in model $\mathbb{G}$ if there is some assignment of values to the parameters in F that yields the probability distribution.

Example 14 *DAG models appear in Figures 2.20 (a) and (b). The values of X, Y, and Z are x1, x2, y1, y2, y3, z1, and z2. The probability distribution contained in the causal network in Figure 2.20 (c) is included in both models, whereas the one in the causal network in Figure 2.20 (d) is included only in the model in Figure 2.20 (b). The reason this latter probability distribution is not included in the model*

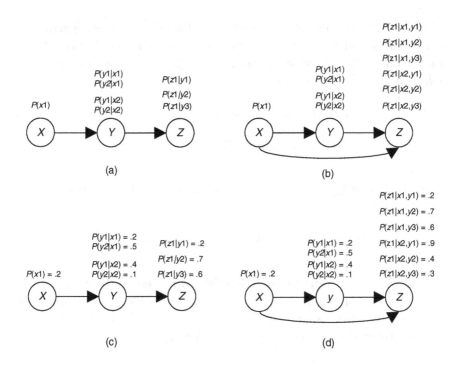

Figure 2.20: DAG models appear in (a) and (b). The probability distribution in the causal network in (c) is included in both models, whereas the one in (d) is included only in the model in (b).

in Figures 2.20 (a) is that any probability distribution included in that model would need to have $I(X, Z|Y)$, and this probability distribution does not have that conditional independency.

The **dimension** of a DAG model is the number of parameters in the model. The dimension of the DAG model in Figures 2.20 (a) is 8, while the dimension of the one in Figures 2.20 (b) is 11.

We now have the following definition concerning scoring criteria:

Definition 1 Let $data_n$ be a set of values (data) of a set of n mutually independent random vectors, each with probability distribution P, and let P_n be the probability function determined by the joint distribution *of the n random vectors. Furthermore, let score be a scoring criterion over some set of DAG models for the random variables that constitute each vector. We say score is* **consistent** *for the set of models if the following two properties hold:*

1. *If $\mathbb{G}_1$ includes P and $\mathbb{G}_2$ does not, then*

$$\lim_{n \to \infty} P_n \left(score(\mathsf{data}_n, \mathbb{G}_1) > score(\mathsf{data}_n, \mathbb{G}_2) \right) = 1.$$

2. *If $\mathbb{G}_1$ and $\mathbb{G}_2$ both include P and $\mathbb{G}_1$ has smaller dimension than $\mathbb{G}_2$, then*

$$\lim_{n \to \infty} P_n \left(score(\mathsf{data}_n, \mathbb{G}_1) > score(\mathsf{data}_n, \mathbb{G}_2) \right) = 1.$$

We call P the **generative distribution**. The limit, as the size of the data set approaches infinity, of the probability that a consistent scoring criterion chooses a smallest model that includes P is 1.

The **Bayesian scoring criterion,** which is the

$$P(\mathsf{data}|\mathbb{G})$$

(i.e. it is the probability of the data given the DAG), is a consistent scoring criterion for DAG models whose parameters are discrete. This criterion and its consistency are discussed in detail in [11]. Presently, we just show a few simple examples illustrating the result of applying it.

Example 15 *Suppose $\mathsf{V} = \{X, Y\}$, both variables are binary, and the set of conditional independencies in probability distribution P is*

$$\{I(X, Y)\}.$$

Let the values of X be $x1$ and $x2$ and the values of Y be $y1$ and $y2$. Possible DAG models are shown in Figure 2.21. Both models include P. When the data set is large, the model in (a) should be chosen by the Bayesian scoring criterion because it has smaller dimension.

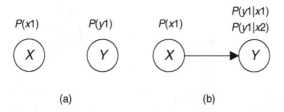

Figure 2.21: Two DAG models.

Example 16 *Suppose* $\mathsf{V} = \{X, Y\}$*, both variables are binary, and the set of conditional independencies in probability distribution P is the empty set*

$$\varnothing.$$

Possible DAG models are shown in Figure 2.21. When the data set is large, the model in (b) should be chosen by the Bayesian scoring criterion because it is the only one that includes P.

A **hidden variable DAG model** is a DAG model augmented with hidden variables. Figure 2.22 (b) shows a hidden variable DAG model. The variables that are not hidden are called **observables**. It has not been proven whether, in general, the Bayesian scoring criterion is consistent when hidden variable DAG models are also considered. However, Rusakov and Geiger [14] proved it is consistent in the case of **naive hidden variable DAG models**, which are models such that there is a single hidden variable H, all observables are children of H, and there are no edges between any observables.

Example 17 *Suppose* $\mathsf{V} = \{X, Y\}$*, both variables have three possible values, and the set of conditional independencies in probability distribution P is the empty set*

$$\varnothing.$$

Let $\mathbb{G}$ *be the DAG model in Figure 2.22 (a) and* $\mathbb{G}_H$ *be the naive hidden variable DAG model in Figure 2.22 (b). Although* $\mathbb{G}_H$ *appears larger, it is actually smaller because it has fewer effective parameters.*

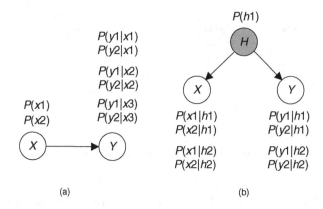

Figure 2.22: A DAG model and a hidden variable DAG model.

This is discussed rigorously in [11]. The following is an intuitive explanation. Since $\mathbb{G}$ includes all joint probability distributions of these variables, $\mathbb{G}$ includes every distribution that $\mathbb{G}_H$ includes. However, $\mathbb{G}_H$ only includes distributions that can be represented by urn problems in which there is a division of the objects into two sets such that X and Y are independent in each set. For example, $\mathbb{G}_H$ includes the joint distribution of the value (1, 2, or 3) and shape (circle, square, or arrow) of the objects Figure 2.23 (a) because value and shape are independent given either the set of black objects or the set of white objects. However, $\mathbb{G}_H$ does not include the joint distribution of the value and shape of the objects in Figure 2.23 (b) because there is no division of the objects into two sets that value and shape are independent in each set.

As discussed in [11], in the space consisting of all possible assignments of values to the parameters of model $\mathbb{G}$ in Figure 2.22 (a), the subset, whose probability distributions are included in hidden variable model $\mathbb{G}_H$ in Figure 2.22 (b), has Lebesgue measure zero. This means that if we assign arbitrary values to the parameters in $\mathbb{G}$, we can be almost certain the resultant probability distribution is not included in $\mathbb{G}_H$.

Consider now the following experiment:

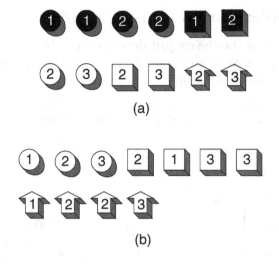

(a)

(b)

Figure 2.23: Value and shape are independent given color in (a). There is no division of the objects in (b) into two sets which renders value and shape independent.

1. Randomly choose either $\mathbb{G}$ or $\mathbb{G}_H$.

2. Randomly assign parameter values to the model chosen.

3. Generate a large amount of data.

4. Score the DAGs based on the data using the Bayesian scoring criterion.

When $\mathbb{G}$ is chosen, almost certainly the probability distribution will not be included in $\mathbb{G}_H$. So with high probability $\mathbb{G}$ will score higher. When $\mathbb{G}_H$ is chosen, with high probability $\mathbb{G}_H$ will score higher because it has smaller dimension than $\mathbb{G}$. So from the data alone we can become very confident as to which DAG model was randomly chosen.

2.3.2   Application to Causal Learning

Next we show how the theory just developed can be applied to causal learning. Suppose some large population is distributed according to the data in the following table:

Case	Sex	Height (inches)	Wage ($)
1	female	64	30,000
2	female	64	30,000
3	female	64	40,000
4	female	64	40,000
5	female	68	30,000
6	female	68	40,000
7	male	64	40,000
8	male	64	50,000
9	male	68	40,000
10	male	68	50,000
11	male	70	40,000
12	male	70	50,000

The random variables *Sex*, *Height*, *Wage* have the same joint probability distribution as the random variables *Color*, *Shape*, and *Value* in Figure 2.23 (a) when we make the following associations:

black/female,   *white/male*,   *circle/64*,   *square/68*,

arrow/70,   *1/30,000   2/40,000*,   *3/50,000*.

Suppose now we only observe and collect data on height and wage. Sex is then a hidden variable in the sense that it renders the observed variables independent. If we only looked for correlation, we would find height and wage are correlated and perhaps conclude height has a causal effect on wage. However, if we score the models $\mathbb{G}$ and $\mathbb{G}_H$ in Figure 2.22, $\mathbb{G}_H$ will most probably win because it has smaller dimension than $\mathbb{G}$. We can then conclude that possibly there is a hidden common cause.

We said 'possibly' because there are a number of caveats when concluding causation from data on two variables. They are as follows:

1. The hidden variable DAG models $X \rightarrow H \rightarrow Y$ and $X \leftarrow H \leftarrow Y$ have the same score as $X \leftarrow H \rightarrow Y$. So we may have a hidden intermediate cause instead of a hidden common cause.

2. In real applications features like height and wage are continuous. We create discrete values by imposing cutoff points. With different cutoff points we may not create a division that renders wage and height independent.

3. Similar caution must be used if the DAG model wins and we want to conclude X causes Y or Y causes X. That is,

    (a) Different cutoff points may result in a division of the objects that renders them independent, which means the hidden variable model would win.

    (b) If there is a hidden common cause, it may be modeled better with a hidden variable that has a larger space. Clearly, if we increase the space size of H sufficiently the hidden variable model will have the same size as the DAG model.

    (c) Selection bias may be present. For example, if X causes Z and Y causes Z, and we are observing a population in which all members have $Z = z'$, then the observed distribution is $P(x, y|z') = P(y|x, z')P(x|z')$, which means the observed distribution is included in the DAG $X \rightarrow Y$.

We conclude that data on two variables can only give us an indication as to what may be going on causally. For example, in the example involving sex, height, and wage, the data can inform us that it seems there may be a binary hidden common cause. Given this indication, we can then investigate the situation further by searching for one.

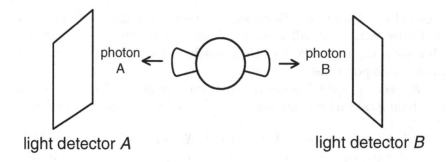

Figure 2.24: Dual photons are emitted in opposite directions.

2.3.3   Application to Quantum Mechanics

The theory of quantum mechanics entails that not all physical observables can be simultaneously known with unlimited precision, even in principle. For example, you can place a subatomic particle so its position is well-defined, but then you cannot determine its momentum. If the momentum is well-defined, you cannot determine its position. It's not just a human measurement problem, but rather that precise values cannot simultaneously be assigned to both momentum and position in the mathematics of quantum mechanics. The assignment of a precise value to momentum determines a unique probability density function of position and vice versa.

The EPR Paradox

According to Einstein, Podolsky, and Rosen [5], a 'complete theory' would be able to represent the simultaneous existence of any properties that do simultaneously exist. They set out to show that properties like momentum and position do simultaneously exist and that therefore quantum mechanics is an incomplete theory. To that end, they developed a thought experiment, called the **EPR Paradox**, similar to the one we describe next. Their thought experiment involved position and momentum. However, since the actual experiments that were eventually done involved spin, we describe this latter experiment.

Figure 2.24 depicts a situation in which a pair of photons is emitted, and measurements are made an arbitrary distance away, which could be many light years. Each photon has a spin along each of the x-axis, y-axis, and z-axis. If photon A's spin on the x-axis is positive, photon B's must be negative. So by measuring photon A's spin on any axis, we can learn Photon B's spin on that axis. Experiments, in which we measured, for example, both spins on the x-axis, have substantiated that they are opposite of each other. Like position and momentum, the three spins of a photon cannot 'simultaneously exist' in quantum mechanics.

Einstein et al. [5] made the 'locality assumption', which says no instantaneous change in photon B can occur owing to a measurement at photon A. So the value of the spin of photon B must have existed before the measurement at A took place. Since we can do this experiment for each of the spin directions, this is true for all three spins. So the values of the three spins must simultaneously exist. We conclude that the spins must have been determined at the time the split was made, and that the conditions at this time constitute a hidden variable that renders the spins of photon A and B independent. We conclude quantum mechanics is incomplete.

Bell's Inequality

In 1964 John Bell [2] showed that if there is a hidden variable, then the probability distribution of the spins must satisfy Bell's Inequality. In our terminology, this means the hidden variable DAG model only includes probability distributions that satisfy Bell's Inequality. However, a number of experiments have shown that Bell's Inequality is not satisfied [1]. We discuss Bell's Inequality, and these results and their ramifications next.

Theorem 2 *(Bell's Inequality) Let*

    1.   X, Y, and Z be any three properties of entities in some population.

    2. $\#(X, Y)$ denote the number of entities with properties X and Y.

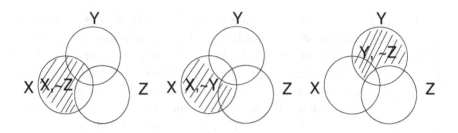

Figure 2.25: A proof of Bell's Inequality.

Then

$$\#(X,\sim Z) \le \#(X,\sim Y) + \#(Y,\sim Z),$$

where '$\sim$' is the Not operator.

Proof. *The proof follows from the Venn diagram in Figure 2.25.*

Now let X_A be a variable whose values are the possible spins of photon A in the x-axis. These values are $x_A \uparrow$ and $x_A \downarrow$. Let similar definitions hold for Y, Z, and B. Then if we emit many photons, according to Bell's Inequality,

$$\#(x_A \uparrow, z_A \downarrow) \le \#(x_A \uparrow, y_A \downarrow) + \#(y_A \uparrow, z_A \downarrow).$$

Since B's directions are always opposite to A's directions, this means

$$\#(x_A \uparrow, z_B \uparrow) \le \#(x_A \uparrow, y_B \uparrow) + \#(y_A \uparrow, z_B \uparrow)$$

These values we can actually measure. That is, we do one measurement at each detector. We cannot obtain all measurements simultaneously. However, we can repeat the experiment many times, each time taking one pair of measurements. This has been done quite a few times, and the inequality was found not to hold [1].

Next suppose, as depicted in Figure 2.26, we have a hidden variable generating six spin values according to some probability distribution of the six spin variables, and this distribution has the marginal distributions

$$P(x_A, z_B), \quad P(x_A, y_B), \quad P(y_A, z_B).$$

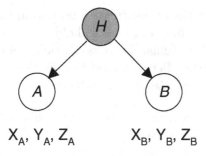

Figure 2.26: A hidden variable is generating values of the six spin variables.

Then if we have n repetitions of each measurement with n large,

$$\#(x_A \uparrow, z_B \uparrow) \approx P(x_A \uparrow, z_B \uparrow) \times n$$
$$\#(x_A \uparrow, y_B \uparrow) \approx P(x_A \uparrow, y_B \uparrow) \times n$$
$$\#(y_A \uparrow, z_B \uparrow) \approx P(y_A \uparrow, z_B \uparrow) \times n$$

and these numbers would have to obey Bell's Inequality. Since, as noted above, the observed numbers do not obey Bell's Inequality, we conclude the observed distributions are not marginals of distributions that are included in the hidden variable DAG model in Figure 2.26. Note that the observed distributions cannot be marginals of any distribution of the six variables. So we also cannot suppose that, at the time of emission, a probability distribution of the six variables is being generated by photon A obtaining three spin values and these values then causing photon B to have three values. A similar statement holds for selection bias.

Note that it may seem that we have 'cheated' quantum mechanics. That is, when we measure x_A and z_B, we learn the values of x_A and z_A (since z_A must be in the opposite direction of z_B). But quantum mechanics says we can't know both those values simultaneously. The following explanation clears this matter up:

This was one of Schrödinger's first reactions to EPR in 1935. The problem is that when you measure x_A on one

> side and z_B on the other, there is no reason to think that the conservation law continues to hold, since you will have disturbed both systems. There is a little proof that it can't hold due to Asher Peres, buried in a little paper of mine.
> - Arthur Fine (private correspondence).

That is, when we measure x_A we disturb photon A; so its z_A value need no longer be the inverse of z_B.

In spite of this disturbance, the hidden variable model is still refuted. That is, if values of all six spin variables had been generated according to some probability distribution before the measurements, then the value of z_B would have been the inverse of the value of z_A right before the measurements. This means the (accurate) measured value of z_B would be the inverse of the value of z_A before the measurements.

So experimental results show that a hidden variable is not possible. Given this, what could explain the correlation between the measurements at the two detectors? Some explanations follow:

1. The measurement at detector A is 'causing' the values of the spins at detector B from many miles away and faster than the speed of light.

2. If you don't insist the spins of the two photons are separates variables, then you don't have any problem with superluminal causation or the violation of independence conditions. That is, although we thought we had two separate properties, we don't. Not only can't we figure out a way to vary them independently, but we have good theoretical reason for thinking we can't. In other words, we have a single variable (whose value are the three spin directions). The odd thing is that this property is nonlocal. That is, the measurement applies to a small region over here, and a small region over there, and nowhere in between - Charles Twardy.

3. We measure values for the left photon, and then the particle travels backwards in time to the source, collides with the source, which then emits the right photon. - Phil Dowe.

If we accept a simple manipulation definition of causation, there is no problem with the first and second explanations above. The two explanations are actually about the same because, in general, in a causal network a deterministic relationship between two variables can be modeled as a single variable. These explanations become difficult only when we insist that causation involves contact, material transfer, etc. Because our macro experiences, which gave rise to the concept of causality, have all involved physical contact, it seems our intuition requires it. However, if we had many experiences with manipulation similar to the ones at the quantum mechanical level, we may have different intuition. Finally, there is an epistemological explanation. Just as humans have the capability to view and therefore model more of reality than the amoeba, perhaps some more advanced creature could better view and model this result and thereby have a better intuition for it than us.

References

1. Aspect A, Grangier P, and Roger G (1982) Experimental Realization of Einstein-Podolsky-Rosen-Bohm Gedanken Experiment: A New Violation of Bell's Inequality, *Physical Review Letters*, 49 #2.

2. Bell JS (1964) On the Einstein-Podolsky-Rosen paradox, *Physics*, 1.

3. Christensen R (1990) *Log-Linear Models*, Springer-Verlag, New York.

4. Druzdzel MJ and Glymour C (1999) Causal Inference from Databases: Why Universities Lose Students, in Glymour C, and Cooper GF (eds.) *Computation, Causation, and Discovery*, AAAI Press, Menlo Park, California.

5. Einstein A, Podolsky B, and Rosen N (1935) Can Quantum-Mechanical Description of Physical Reality be Consdiered Complete, *Physical Review*, 47.

6. Heckerman D, Meek C, and Cooper G (1999) A Bayesian Approach to Causal Discovery, in Glymour C, and Cooper GF (eds.) *Computation, Causation, and Discovery*, AAAI Press, Menlo Park, California.

7. Lugg JA, Raifer J, and González CNF (1995) Dehydrotestosterone is the Active Androgen in the Maintenance of Nitric Oxide-Mediated Penile Erection in the Rat, *Endocrinology*, 136(4).

8. McClennan KJ and Markham A (1999) Finasteride: A review of its Use in Male Pattern Baldness, *Drugs*, 57(1).

9. Meek C (1995) Causal Influence and Causal Explanation with Background Knowledge, in Besnard P and Hanks S (eds.) *Uncertainty in Artificial Intelligence; Proceedings of the Eleventh Conference*, Morgan Kaufmann, San Mateo, California.

10. Neapolitan RE (1990) *Probabilistic Reasoning in Expert Systems*, Wiley, New York.

11. Neapolitan RE (2003) *Learning Bayesian Networks*, Prentice Hall, Upper Saddle River, New Jersey.

12. Neapolitan RE and Morris S (2004) Probabilistic Modeling Using Bayesian Networks, in Kaplan D (ed.) *Handbook of Quantitative Methodology in the Social Sciences*, Sage, Thousand Oaks, California.

13. Pearl J (2000) *Causality: Models, Reasoning, and Inference*, Cambridge University Press, Cambridge, United Kingdom.

14. Rusakov D, and Geiger D (2002) Bayesian Model Selection for Naive Bayes Models, in Darwiche A, and Friedman N (eds.) *Uncertainty in Artificial Intelligence; Proceedings of the Eighteenth Conference*, Morgan Kaufmann, San Mateo, California.

15. Scarville J, Button SB, Edwards JE, Lancaster AR, and Elig TW (1996) Armed Forces 1996 Equal Opportunity Survey, Defense Manpower Data Center, Arlington, VA. DMDC Report No. 97-0279.

16. Scheines R, Spirtes P, Glymour C, and Meek C (1994) *Tetrad II: User Manual*, Lawrence Erlbaum, Hillsdale, New Jersery.

17. Spirtes P, Glymour C, and Scheines E (1993, 2000) *Causation, Prediction, and Search*, Springer-Verlag, New York; 2nd ed.: MIT Press, Cambridge, Massachusetts.

18. Stangor C, Swim JK, Van Allen KL, and Sechrist GB (2002) Reporting Discrimination in Public and Private Contexts, *Journal of Personality and Social Psychology*, 82.

3 Learning Based Programming

Dan Roth

Dept. of Computer Science , University of Illinois , Urbana, IL 61801,
USA danr@cs.uiuc.edu

Abstract

A significant amount of the software written today, and more so in the
future, interacts with naturally occurring data — text, speech, images and
video, streams of financial data, biological sequences — and needs to
reason with respect to concepts that are complex and often cannot be
written explicitly in terms of the raw data observed. Developing these
systems requires software that is centered around a semantic level
interaction model, made possible via trainable components that support
abstractions over real world observations.

Today's programming paradigms and the corresponding programming
languages, though, are not conducive for that goal. Conventional
programming languages rely on a programmer to explicitly define all the
concepts and relations involved. On the other hand, in order to write
programs that deal with naturally-occurring data, that is highly variable
and ambiguous at the measurement level, one needs to develop a new
programming model, in which some of the variables, concepts and
relations may not be known at programming time, may be defined only in
a data driven way, or may not be unambiguously defined without relying
on other concepts acquired this way.

In Learning Based Programming (LBP), we propose a programming
model that supports interaction with domain elements at a semantic level.
LBP addresses key issues that will facilitate the development of systems
that interact with real-world data at that level by (1) allowing the
programmer to *name* abstractions over domain elements and information
sources – defined implicitly in observed data, (2) allowing a programmer
to interact with named abstractions (3) supporting seamless incorporation
of trainable components into the program, (4) providing a level of
inference over trainable components to support combining sources and
decisions in ways that respect domain's or application's constraints, and
(5) a compilation process that turns a data-dependent high level program
into an explicit program, once data is observed.

D. Roth: *Learning Based Programming*, StudFuzz **194**, 73–95 (2006)
www.springerlink.com   © Springer-Verlag Berlin Heidelberg 2006

This chapter describes preliminary work towards the design of such a language, presents some of the theoretical foundations for it and outlines a first generation implementation of a Learning based Programming language, along with some examples.

3.1 Introduction

...these machines are intended to carry out any operations which could be done by a human computer ... supposed to be following fixed rules. ... We may suppose that these rules are supplied in a book which is altered whenever he is put on to a new job....

A. M. Turing, Computing Machinery and Intelligence, 1950 B.

The fundamental question that led Turing and others to the development of the theory of computation and of conventional programming languages is the attempt to model and understand the limitation of mechanical computation. Programming languages were then developed to simplify the process of setting explicit rules of computation that digital machines can follow. In the last two decades we have seen the emergence of the field of Machine Learning that studies how concepts can be defined and represented without the need for explicit programming, but rather by being presented with previous experiences. Quoting Leslie Valiant, the theory of machine learning attempts to study models of "what can be learned just as computability theory does on what can be computed".

The goal of Learning Based Programming is to develop the programming paradigm and language that can support machine learning centered systems. We view this as a necessary step in facilitating the development of computer programs that interact with and make inferences with respect to naturally occurring data.

Consider writing a computer program that controls an automatic assistant for analyzing a surveillance tape; the programmer wishes to reason with respect to concepts such as *indoor and outdoor scenes*, the identification of *humans* in the image, *gender* and *style of clothing*, recognizing *known people* (perhaps based on typical gestures or movements) etc. Each of these concepts depend on a large number of hierarchical decisions with respect to the observed input. These might include recognizing the concepts mentioned above, tracking a moving body, associating video with audio, etc. These, in turn, depend in intricate ways on even lower level decisions such as identifying trees or other objects, that are also very involved in terms of the raw input presented. Given that no two inputs observed by this program are ever the *same*, it seems impossible to explicitly write down the dependencies among all factors that may potentially contribute to a decision, or a Turing-like list of definitions and rules the program can follow in its processing. Consequently, a lot of research in the last few years focuses on developing machine learning methods that use observed data to *learn* functions that can reliably recognize these lower level concepts.

Today, developing systems like the one mentioned above resembles programming in *machine language*, as opposed to using an advanced programming language. There is no programming paradigm that supports embedded trainable modules and that allows the development of programs that interact with real worlds data and can abstract away most of the lower level details – including defining features and training classifiers to represent low level predicates – and reason, perhaps probabilistically, at the right level of abstraction. Today, in fact, writing a program for each of the low level tasks mentioned above is a respectable Ph.D. thesis.

Other forms of intelligent access to information can serve as examples for the type of processes future programs will have to support. Consider a program that interacts with free form natural langauge questions presented by users attempting to extract information from some news source, the web or even a knowledge base. E.g., "at what elevation did the Columbia explosion occur?" or, more mundane questions like "Is there public transportation from Campus to Urbana on Sundays?" Naturally, interpreting a natural language question to the extent that it can yield a reliable mapping into a knowledge base query or support extracting an answer from free form documents is a challenging task that requires resolving context-sensitive ambiguities at several levels. A lot of research is done on developing machine learning based approaches that deal with specific context sensitive ambiguities, ranging from predicting the part-of-speech tag of a word in the context of a given sentence, resolving the sense of a word in a given context, resolving prepositional phrase attachments, inducing a syntactic and semantic parse of a sentence, handling a query presented in the context of a previous one, tolerating user's input errors (e.g. typing our instead of out, now instead of know), etc. Each of these specific tasks is a difficult undertaking that qualifies for a Ph.D. thesis. However, there is no programming paradigm that allows people to reason at the desired task level, define and learn the appropriate concepts, chain experience based decisions, and reason with them.

While the development of machine learning libraries is an important stage towards facilitating broader use of software systems that have learning components, algorithmic libraries are only one component in Learning Based Programming. LBP focuses on providing a programmer a way to interact with real world data at the appropriate level of abstraction and reason with induced properties of real world domain elements. Starting from *naming* the key primitive elements that can be *sensed* in a domain, to supporting hierarchical naming via learning operators to enabling reasoning with respect to domain elements and their properties by defining constraints among variables in the domain.

3.2 Learning Based Programming

Learning Based Programming (LBP) is a programming paradigm that extends conventional programming languages by allowing a programmer to write programs in which some of the variables are not explicitly defined in the program. Instead, these variables can be *named* by the programmer that may use language constructs to define them as trainable components. That is, these variables are defined, possibly recursively, as outcomes of trainable components operating on data sources supplied to the program. Conventional programming languages allow the design and implementation of large scale software systems and rely on a programmer to explicitly define all the concepts and relations involved, often hierarchically. In LBP we develop a programming model that supports building large scale systems in which some components cannot be explicitly defined by a programmer. Realizing that some of the variables, concepts and relations may not have an explicit representation in terms of the observed data, or may not be defined unambiguously without relying on other higher level concepts, LBP provides mechanisms that allow seamless programming using data-defined concepts.

Learning Based Programming allows the programming of complex systems whose behaviors depend on naturally occurring data and that require reasoning about data and concepts in ways that are hard, if not impossible, to write explicitly. The programmer can reason using high level concepts without the need to explicitly define all the variables they might depend on, or the functional dependencies among them. It supports reasoning in terms of the information sources that might contribute to decisions; exactly what variables are extracted from these information sources and how to combine them to define other variables is determined in a data-driven way, via a learning operator the operation of which is abstracted away from the programmer. The key abstraction is that of *naming*, which allows for re-representation of domain elements in terms other concepts, low level *sensors* or higher level concepts that are functions of them. Naming is supported in LBP by allowing programming in terms of data-defined, trainable components. This makes an LBP program, formally, a family of programs; yet, an explicit program is eventually generated by a compilation process that takes as additional input the data observed by the program.

The rest of this article describes the technical issues involved in developing this programming paradigm. The technical description focuses on the knowledge representations underlying LBP, since this is a key issue in developing the language.

3.3 The LBP Programming Model

A computer program can be viewed as a system of definitions. Programming languages are designed to express definitions of variables in terms of other variables, and allow the manipulations of them so that they represent some meaningful concepts, relations or actions in a domain. In traditional programming systems the programmer is responsible for setting up these definitions. Some may be very complicated and may require the definition of a hierarchy of functions. In LBP, a *naming* process allows the declaration of variables in terms of other variables without setting up the explicit functional dependencies; explicit dependencies are being set up only at a later stage, via interactions with data sources, using a learning process that is abstracted away from the programmer. LBP provides the formalism and an implementation that allows a programmer to set up definitions only by declaring the target concepts (the variables being defined) and the information sources a target concept might depend on, without specifying exactly how. Variables defined this way might have interacting definitions (in the sense that x is a data-defined variable that depends on y, and y is a data-defined variable that depends on x), and there might be some constraints on simultaneous values they can take (e.g., Boolean variables x and y cannot take the same value).

Associated with data sources supplied to an LBP program is a well defined notion of a *sensor*. A sensor is a functional (defined explicitly below) that provides the abstraction required for an LBP program to interact with data. A given domain element comes with a set of sensors that encode the information a program may know about this element. That is, sensors "understand" the raw representation of the data source at some basic level, and provide a mechanism to define and represent more abstract concepts with respect to these elements. For example, for programs that deal with textual input, there might exist basic predicates that *sense* a "word", the relation "before" between words (representing that the word u is before word w in the sentence), and a "punctuation mark", in the input. Other sensors might also be supplied, that provide the program with more knowledge, such as "part-of-speech tag of a word", or a sensor that identifies upper case letters. Primitive sensors might be provided with the data, or programmed to reflect an understanding of the input data format. All other variables will be defined in terms of the output of these sensors, some explicitly and some just *named*, and defined as trainable variables.

Unlike conventional programming models, the notion of an *interaction* is thus central in LBP. An LBP program is data driven. That is, some of the variables participating in the declaration of a new variable, as well as the exact ways new variables depend on other variables, are determined during a data-driven compilation process, which is implemented via machine learning programs and inference algorithms. An LBP program is thus a set of possible programs. The specific element in this set of possible programs that will perform the computation is determined either in a *data driven compilation* stage or at run-time. The notion of data-driven compilation is fundamental to LBP, which has, conceptually, two compilation stages. The first is a machine learning based compilation in which the program interacts with data sources. In this stage, the LBP code, which may contain data-defined variables, is converted into a "conventional" program – although many of the predicates may have complex definitions the programmer may not fully understand (and need not look at). This is done by "running" the machine learning algorithm and inference algorithms on the LBP code and the data pointed to from it. The second compilation stage is the conventional stage of converting high level code to machine code.

Specifically, an LBP program extends traditional programming in the following ways:

- It allows to define *types* of variables, that may give rise to (potentially infinitely) many variables, the exact name of which is determined in a data-driven way.
- Variables (and types of variables) are *named* as abstractions over domain elements; they are defined as the outcome of trainable components and their explicit representation is learned in terms of the sensory input.
- It supports a well defined notion of *interaction* with data sources, via sensor mechanism that abstracts the raw information available in the data source.
- It allows the definition of *sets* of data-defined and possibly mutually constrained variables; these take values that are optimized globally via inference procedures.
- Data-driven compilation converts an LBP program into a conventional program by first defining the optimal set of

learningtasks required, and a training stage that results in the generation of an explicit program.

Although LBP is formalized in the traditional manner, in the sense that it is given a syntax and operational semantics, we focus here on describing an abstract view of some of the constructs, without providing the syntax of our implementation, so that the general view of the language and programming with it is clear.

3.3.1. Knowledge Representations for LBP

The key abstraction in LBP is that of a *Relational Generation Function (RGF)*. This is a notion that allows LBP to treat both explicitly defined variables and (types of) variables that are data-defined via learning algorithms, as basic, interchangeable building blocks. In order to define RGFs we need to introduce two of the basic representational constructs in LBP – *relational variables* and a *structural instance space*. An LBP program can be viewed as a program for manipulating, augmenting and performing inferences with respect to elements in the structural instance space. The augmentation can be explicit (changing the instance space) or implicit, and amounts to re-representation of (properties of) domain elements as functions of other (properties of) domain elements, typically done via learning operators incorporated into the program. From the programmer's perspective, though, handling the knowledge representation is done in the same way regardless of whether elements are readily available in the input (via sensors; see below), explicitly defined, or named via learning or inference operators.

A structured instance is a multi-labeled graph based representation of data; nodes are place holders for individuals in the domain; edges represent relations among individuals; and labels (possibly multi labels) on nodes and edges represent properties of these. Raw data elements the program interacts with are viewed by the program via its associated *sensors* (to be defined below), that reveal the information available in the input. The knowledge representation described below, which underlies LBP, draws on our previous work for knowledge representations that support learning

For example, given a natural language *sentence* in the input, the readily available information might be the words and their order (via a word sensor and a before (a,b) sensor). The programmer might want to reason with respect to higher level predicates such as the syntactic category of the word (e.g., noun, verb), the semantic sense of a word, or its role in the sentence (e.g., subject, object). Reasoning at that level constitutes augmentation of the structured instance by adding labels, nodes (e.g., representing noun phrases) and edges to the graph. LBP allows the programmer to define these and reason with the higher level predicates. This *naming* operation can be viewed as re-representing the input in terms of higher level concepts, the definition of which in terms of the readily available sensors cannot be explicitly given by the programmer[1]. Similarly, a *document* in the input might be represented as a list of words (same sensors as above), but the programmer might want to reason about it also as a list of *sentences*, or as a structured document with a *title* and *paragraphs*[2].

We define below a collection of "relational" variables, termed so to emphasize that they can actually stand for relational (quantified) entities. The relational variables are all formulae in a relational language R that we now define[3]. We then describe how a programmer can initiate a data driven generation of these variables, rather than explicitly define all of them, how they can be manipulated, and how they can be defined in terms of others. For most of the discussion we treat relational variables as Boolean variables, taking only values in $\{0,1\}$. It will be clear that these variables can also take real-values (in the range $[0,1]$).

The data sources the program interacts with constitute the *domain* of the relational language R. A domain consists of elements and relations among these elements. Predicates in the domain either describe the relation between an element and its attributes, or the relation between two elements.

Definition 1. (Domain) *A domain* $D = \langle V, E \rangle$ *consists of a set of typed elements,* V, *and a set of binary relations* E *between elements in* V. *An element is associated with some attributes and two elements may have multiple relations between them. When two elements have different sets of attributes, we say that the two elements belong to different types.*

---

[1]Clearly, an explicit definition of the syntactic role of a word in the sentence cannot be given and is likely to depend on the value of the syntactic role of neighboring words; the learning operator will be used to abstract away the explicit definition, acquired in the data-driven compilation process.

[2]Even segmenting a document into sentences is a non trivial and context sensitive task, as a large number of the "periods" that occur in documents do not represent sentence delimiters. Similarly, identifying the structure of the document, e.g., title, abstract, introduction and other sections is delegated to named variables that are defined in a data driven way.

[3]In recent work we have developed two different, but equivalent ways to represent elements in the language R. We provide here the operational definition, but a well defined syntax and semantics for the language can also be defined via feature description logic [5].

Definition 2. (Instance) *An instance is an interpretation [13] which lists a set of domain elements and the truth values of all instantiations of the predicates on them.*

Each instance can be mapped to a directed graph. In particular, each node represents an element in the domain, and may have multiple labels, and each link (directed edge) denotes the relation that holds between the two connected elements. Multiple edges may exists between two given nodes.

The relational language R is a restricted (function free) first order language for representing knowledge with respect to a domain D. The restrictions on R are applied by limiting the formulae allowed in the language to those that can be evaluated very efficiently on given instances (Def. 2). This is done by (1) defining primitive formulae with a limited scope of quantifiers, and (2) defining general formulae inductively, in terms of primitive formulae, in a restricted way that depends on the relational structures in the domain. The emphasis is on locality with respect to the relational structures that are represented as graphs.

We omit many of the standard FOL definitions and concentrate on the unique characteristics of R (see, e.g., [13] for details). The vocabulary in R consists of constants, variables, predicate symbols, quantifiers, and connectives. Constants and predicate symbols vary for different domains. In particular, for each constant in R, there is an assignment of an element in V. For each k-ary predicate in R, there is an assignment of a mapping from V^k to $\{0,1\}$ ($\{$true, false$\}$).

Primitive formulae in R are defined in the standard way, with the restriction that all formulae have only a single predicate in the scope of each variable. Notice that for primitive formulae in R, the *scope* of a quantifier is always the unique predicate that occurs within the atomic formula. We call a variable-free atomic formula a *proposition* and a quantified atomic formula, a *quantified proposition* [11]. Clearly, for any Boolean function $f : \{0,1\}^n \rightarrow \{0,1\}$ on n variables, if $F_1, F_2, \ldots F_n$ are primitive formulae in R, then $f(F_1, F_2, \ldots, F_n)$ is also a formula. The informal semantics of the quantifiers and connectives is the same as usual. Relational variables in R receive their "truth values" in a data driven way, with respect to an observed instance. Given an instance x, a formula F in R has a unique truth value, *the value of F on x*, defined inductively using the truth values of the predicates in F, and the connectives' semantics.

1. Relation Generation Functions

Writing a formula in R provides one way to define new variables. We view a formula F in R as a *relation* $F : x \to \{0,1\}$, which maps the instance x to its truth value. A formula is *active* in x if it has truth value *true* in this instance. We denote by X the set of all instances, the *instance space*. A formula $F \in R$ is thus a relation over X, and we call it a *relational variable*. From this point we use formula and relation interchangeably.

Example 1. *Consider an instance x represented simply as a collection of nodes, that is, as an unordered collection of place holders with two attributes, "word" and "tag" each. Let the list of words be: he, ball, the, kick, would. Then some active relations on this instance are word(he), word(ball), and tag(DET). In this case we could say that the variable word(he) has value 1 in the instance. If object was another attribute of nodes in the domain, or if there was a way to compute this predicate given the input instance, object(z) would be another variable. It represents an unbound formula that would evaluate to true if a word exists in the input instance which is an object of a verb in it.*

We now define one of the main constructs in LBP. The notion of *relation generating functions* allows an LBP programmer to define a collection of relational variables without writing them explicitly. This is important, in particular, in the situations for which we envision LBP is most useful; that is, over very large (or infinite) domains or in on-line situations where the domain elements are not known in advance, and it is simply impossible to write down all possible variables one may want to define. However, it may be possible for the programmer to define all "types" of variables that might be of interest, which is exactly the aim of formalizing the notion of an RGF. As we will see, RGFs will allow a unified treatment of programmer defined (types) of variables and data-defined (types) of variables, which is a key design goal of LBP.

Definition 3. (Relation Generation Function) *Let X be an enumerable collection of formulae over the instance space X. A relation generation function (RGF) is a mapping $G : X \to 2^X$ that maps an instance $x \in X$ to a set of all elements in X that satisfy $\chi(x) = 1$. If there is no $\chi \in X$ for which $\chi(x) = 1$, $G(x) = \phi$.*

RGFs provide a way to define "types" of relational variables (formulae), or to parameterize over a large space of variables. A concrete variable (or a collection thereof) is generated only when an instance x is presented. Next we describe more ways to do that in LBP.

The family of relation generation functions for R are RGFs whose output are variables (formulae) in R. Those are defined inductively, via a *relational calculus*, just like the definition of the language R. The relational calculus is a calculus of symbols that allows one to inductively compose relation generation functions. Although we do not specify the syntax of the calculus here, it should be clear that this is the way an LBP programmer will introduce his/her knowledge of the problem, by defining types of formulae they think are significant. The alphabet for this calculus consists of (i) basic RGFs, called *sensors* and (ii) a set of connectives. While the connectives are part of the language and are the same for every alphabet, the *sensors* vary from domain to domain.

A sensor is the mechanism via which domain information is extracted. Each data source comes with a set of "sensors" that represent what the program can "sense" in it. For example, humans that read a sentence in natural language "see" a lot in it including, perhaps, its syntactic and semantic analysis and even some association to visual imagery. A program might "sense" only the list of words and their order.

A sensor is the LBP construct used to abstract away the way in which information about an instance has been made available to the program. The information might be readily available in the data, might be easily computed from the data, might be a way to access and external knowledge source that may aid in extracting information from an instance, or might even be a previously learned concept.

An LBP programmer programs sensor definitions that encode the knowledge available about the input representation; that is, they parse the input and read from it the type of information that is readily available.

Definition 4. (Sensor) *A sensor is a relation generation function that maps an instance x into a set of atomic formulae in R. When evaluated on instance x, a sensor s outputs all atomic formulae in its range which are active.*

Once sensors are defined, the relational calculus is a calculus of symbols that allows one to inductively compose relation generation functions with sensors as the primitive elements. The relational calculus allows the inductive generation of new RGFs by applying connectives and quantifiers over existing RGFs (see [4] for details). Several mechanisms are used in the language to define the operations of RGFs and their design. These include (1) a *conditioning* mechanism, that restricts the range of an RGF to formulae in a given set (or those which satisfy a given property), (2) a *focus* mechanism that restricts the domain of an RGF to a specified part of the instance and (3) a naming mechanism that allows an easy manipulation of RGFs and a simple way to define new RGFs in terms of existing ones. The operation of the RGFs is defined inductively starting with the definitions of the sensors, using these mechanisms and standard definitions of connectives. We provide here the definition of the *focus* mechanism, which specifies a subset of elements in an instance on which an RGF can be applied.

Definition 5. *Let E be a set of elements in a given instance x. An RGF r is focused on E if it generates only formulae in its range that are active in x due to elements in E. The focused RGF is denoted by $r[E]$.*

In particular, the focus mechanism is the one which allows the LBP programmer to define segments of the input, which, once re-represented via the RGFs, form the "examples" presented to the learning operator. There are several ways to define a focus set. It can be specified explicitly or described indirectly by using the structure information (i.e., the links) in the instance. For example, when the goal is to define an RGF that describes a property of a single element in the domain, e.g., part-of-speech tag of a word in a text fragment, focus may be defined relative to this single element. When the goal is to define an RGF the describes a property of the whole instance (e.g., distinguish the mutagenicity of a compound), the focus can simply be the whole instance.

The discussion above exemplified that defining RGFs is tightly related to the structure of the domain. We now augment the relational calculus by adding structural operations, which exploit the structural (relational) properties of a domain as expressed by the links. RGFs defined by these structural operations can generate more general formulae that have more interactions between variables but still allow for efficient evaluation and subsumption, due to the graph structure.

2. Structural Instance Space

We would like to support the ability of a programmer to deal with naturally occurring data and interact with it at several levels of abstractions. Interacting with domain elements at levels that are not readily available in the data and reasoning with respect predicates that are (non trivial) functions of the data level predicates, require that learning operators can induce a representation of the desired predicates in terms of the available ones. The goal of the following discussion is to show how this is facilitated within LBP.

LBP makes it possible to define richer RGFs, as functions of sensors and recursively, as functions of named variables. These, in turned, provide a controlled but enriched representation of data and thus supports the induction of data elements properties via the learning operators. While this is a general mechanism within LBP, each use of it can be viewed in an analogous way to a feature space mapping commonly done in machine learning (e.g., via kernels) [6,10]

LBP supports these mappings via the abstraction of data elements the system interacts with into a *structural instance space*. The structural instance space is an encoding of domain elements as "sensed" by the system as graphs; thus, RGFs can operate on instances in a unified way and an LBP programmer can deal with the manipulation and augmentation of the instances and with inference over them.

We mentioned that each instance can be mapped into a directed graph. The following definition formalizes it. In particular, each node represents an element in the domain, and may have multiple attributes, and each link (directed edge) denotes the relation that holds between the two connected elements. Multiple edges may exist between two given nodes.

Definition 6. (Structured Instance) *Let x be an instance in the domain $D = (V, E)$. The structured instance representation of x is a tuple $(V, E_1, E_2, ... E_k)$ where V is a set of typed elements, and $E_i \subseteq E$ is a set of binary relations of a given type. The graph $G_i = (V, E_i)$, is called the i th structure of the instance x and is restricted to be an acyclic graph on V.*

Example 2. (NLP) *A structured instance can correspond to a sentence, with V, the set of words in the sentence and $E = E_1$ describing the linear structure of the sentence. That is, $(v_i, v_j) \in E_1$ iff the word v_i occurs immediately before v_j in the sentence.*

Example 3. (VP) *A structure instance can correspond to a gray level representation of an image. V may be the set of all positions in a 100×100 gray level image and $E = E_1$ describes the adjacency relations top-down and left-right in it. That is, $(v_j, v_k) \in E_1$ iff the pixel v_j is either immediately to the left or immediately above v_k.*

Alternatively, one could consider a subset of V which corresponds to those nodes in V, those which are deemed interesting (e.g.,[1]) and $E = E_2$ would represent specific relations between these that are of interest. It will be clear later that LBP can allow to easily define or induce new attributes that would thus designate a subset of V.

The relational calculus discussed in Sec. 3.1.1 can now be augmented by adding structural operations. These operations exploit the structural properties of the domain as expressed in each graph G_i^s in order to define RGFs, and thereby generate non-atomic formulae that may have special meaning in the domain. Basically, structural operations allow us to construct RGFs that conjunct existing RGFs evaluated at various nodes of the structural instances. A large number of regular expression like operations can be defined over the graphs. For computational reasons, we recommend, and exemplify below, only operations along chains in the graphs of the structured instance. However, we note that multiple edges types can exist simultaneously in the graph, and therefore, while very significant computationally, we believe that this does not impose significant expressivity limitations in practice.

Example 4. (NLP) *Assume that the structured instance consists of, in addition to the linear structure of the sentence (G_1), a graph G_2 encoding functional relations among the words. RGFs can be written that represent the Subject-Verb relations between words, or the Subject-Verb-Object chain.*

Example 5. (VP) *An RGF can be written that defines an edge relation in an image, building on a sensor s producing active relations for pixels with intensity value above 50. More general operations can be defined or induced to indicate whether a collection of nodes form an edge.*

We exemplify the mechanism by defining two structural operators that make use of the chain structure in a graph; both are types of "collocation" operators.

Definition 7. (collocation)   *Let $s_1,...,s_k$ be RGFs for R, g a chain-structured subgraph in a given domain $D = (V,E)$. $colloc_g(s_1,...,s_k)$ is a restricted conjunctive operator that is evaluated on a chain of length k in* *g. Specifically, let* $v_1,...,v_k \in V$ *be a chain in g. The formulae generated by* $colloc_g(s_1,...,s_k)$ *are those generated by* $s_1[v_1] \& s_2[v_2] \& ... \& s_k[v_k]$ *where*

1. *by* $s_j[v_j]$ *we mean here the RGF* s_j *is focused to* v_j, *and*

2. *the & operator means that formulae in the output of (s&r) are active formulae of the form F^G, where F is in the range of s and G is in the range of r. This is needed since each RGF in the conjunction may produce more than one formula.*

The labels of links can be chosen to be part of the generated features if the user thinks the information could facilitate learning.

Example 6. *When applied with respect to the graph g which represents the linear structure of a sentence,* $colloc_g$ *generates formulae that corresponds to n-grams. E.g., given the fragment "Dr John Smith", RGF colloc(word, word) extracts the bigrams word(Dr)-word(John) and word(John)-word(Smith). When the labels on the links are shown, the features become word(Dr)-before-word(John) and word(John)-before-word(Smith). If the linguistic structure is given instead, features like word(John)-SubjectOf-word(builds) may be generated. See [7] for more examples.*

Similarly to $colloc_g$, one can define a *sparse* collocation as follows:

Definition 8. (sparse collocation) *Let* $s_1,...,s_k$ *be RGFs for R, g a chain structured subgraph in a given domain* $D = (V,E)$. $scolloc_g(s_1,...,s_k)$ *is a restricted conjunctive operator that is evaluated on a chain of length n in g. Specifically, let* $v_1,v_2,...,v_n \in V$ *be a chain in g. For each subset* $v_{i1},v_{i2},...,v_{ik}$, *where* $i_j < i_1$ *when* $j < 1$, *all the formulae:* $s_1[v_{i1}] \& s_2[v_{i2}] \& ... \& s_k[v_{ik}]$ *are generated.*

Example 7. *Given the fragment "Dr John Smith", the features generated by RGF scolloc(word,word) are word(Dr)-word(John), word(Dr)-word(Smith), and word(John)-word(Smith).*

Notice that while primitive formulae in R have a single predicate in the scope of each quantifier, the structural properties provide a way to go beyond that, but only in a restricted way dictated by the RGF definition and the defined graph structure of the domain elements. Structural operations allow us to define RGFs that constrain formulae evaluated on different objects without incurring the cost usually associated with enlarging the scope of free variables. This is done by enlarging the scope only as required by the structure of the instance.

3.3.2. Interaction

An LBP program interacts with its environment via data structures that are *structured instances* along with a set S of *sensors*. A structured instance is an implementation of Def. 6. It is a graph which consists of nodes (place holders) labeled with attributes – a list of *properties* (e.g., word, tag, intensity) that are predicates that hold in this instance. The structure of the instance is described by labeled edges, where the labels represent relations between the corresponding nodes. Multiple edges may exist between nodes.

In order for a program to process a domain element it needs to have an understanding of the domain and how it is represented. There is a need to "parse" the input's syntactically at some level. This is one aspect of what is represented by sensors.

Along with observations which represent, for example, a sentence or an image, the program has a set of sensors which can operate on the observation and provide the information the program "sees" in the input. Sensor use their understanding of the syntactic representation of the input to extract information directly from the input – word, tag, word order or intensity value – but can also utilize outside knowledge in processing the input; for example, a vowel sensor (which outputs an active variable if its focus word starts with a vowel), needs to use some information to determine its output. An IS-A sensor may need to access an external data structure such as wordnet in order to return an active value (e.g., when focused on "desk" it might return "furniture" among the active variables). An LBP programmer thus assumes that domain elements come with parsing information – a set of sensors – or it needs to code these in order to build on it in further processing of the data. Via the sensor mechanism LBP supports mapping real world data (text, images, biological sequences) into structured instances. From that point, programming proceeds as usual, and may consist of manipulating conventional variables, defining new RGFs as functions of sensors, evaluating them on observations, or defining new RGFs without supplying explicit definition for how they are computed, which we discuss next.

3.3.3. Learning Operators in LBP

The design of the learning operators is one of the most important components in LBP and we outline some of the basic ideas on that below. We deliberately disregard a large number of related theoretical issues that are research issues in machine learning and learning theory. The goal of LBP is to build on the success of research in these areas and provide a vehicle for developing systems that use it. The design allows for easy ways to incorporate newly developed learning algorithms and even learning protocols into the language. Conceptually, we would like to think of the learning operators in LBP as the multiplication operation in conventional programming languages in that the specific details of the algorithms may differ according to the type of the operands and are typically hidden from the programmer. At the same time, we expect that the availability of the LBP vehicle will motivate the study of new learning algorithms and protocols as well as open up possibilities for experimental studies in learning that are difficult to pursue today.

The knowledge representations discussed before provide a way to generate and evaluate intermediate representations efficiently, given an observation. An RGF maps an observation into a set of variables, and is thus a *set function*. When defined using the operator $Ł$ below, an RGF would be used mostly as a multi-valued function.

Definition 9. *Let X be the instance space, $C = \{c_1, ... c_k\}$ a discrete set of labels. A set-function $G : X \to C$ maps an instance $x \in X$ to a set $c \subseteq C$ of labels. G is a multi-valued function if $|c| = 1$.*

In practice, our implementation allows G to return real values and, once normalized, can be viewed as supplying a probability distribution over the set elements. This is sometimes important for the inference capabilities discussed later.

An operator Ex is defined and used to evaluate an RGF (or a collection thereof). Given an RGF(s) and an instance as input, it generates a list of all formulae specified by the RGFs that are active in the instance. This operator can be viewed as converting domain elements, typically, using specifically focused RGFs, into examples that conventional learning algorithm (hidden inside the learning operator) can make use of. Ex has several parameters that we do not describe here, which allow, for example, to associate a weight with the produced formulae. The resulting representation is an extension of the infinite attribute model [2], that has been used in the SNoW learning architecture [3]. This representation does not have an explicit notion of a label. Any of the variables (features) in the example can be treated as a label. If needed, the programmer can define an RGF to focus on part of the observation and use it as a label or can generate an RGF specifically to label an observation using external information.

The goal of the knowledge representation and the formalism discussed so far is to enable the next step, namely the process of *learning* representations of RGFs from observations. The $Ł$ operator learns a definition of a multi-valued RGF in terms of other variables. It receives as input a collection $r_1,...r_k$ of RGFs, along with a collection of structured instances and a set of values (names of variables), $T = \{t_1,...,t_k\}$, that determines its range. Elements in the set T are called the target variables. Notice that a new RGF is defined to be a function of the *variables* produced by RGFs in its domain, but the variables themselves need not be specified; only "types" of variables are specified, by listing the RGFs.

$$T \doteq Ł(T, r_1, r_2, ...r_k, instances).$$

This expression defines a new RGF that has the variables' names in T as its possible values. Which of these values will be active when T is evaluated on a future instance is not specified directly by the programmer, and will be learned by the operator $Ł$.

$Ł$ is implemented via a multi-class learner (e.g., [?,9]). When computing $Ł$, the RGFs $r_1,...r_k$ are first evaluated on each instance to generate an example; the example is then used be $Ł$ to update its representation for all the elements in T as a function of the variables that are currently active in the example. The specific implementation of the operator $Ł$ is not important at this level. The key functional requirement that any implementation of $Ł$ needs to satisfy is that it uses the infinite attribute model (otherwise, it needs to know ahead of time the number of variables that may occur in the input.) Thus, examples are represented simply as lists of variables and the operator $Ł$ treats each of the targets as an autonomous learning problem (an inference operator, below, will handle cases of *sets* of variables that receive values together.) It is the responsibility of the programmer to make sure that the learning problem set up by defining T as above is well defined. Once so defined, T is a well defined RGF that can be evaluated by the operator Ex. Moreover, each of the variables in the target set T is an RGF in itself, only that it has only a single variable in its range.

The operator operator $Ł$ has several modes which we do not describe here. We note, though, that from the point of view of the operator $Ł$, learning in LBP is always supervised. However, this need not be the case at the application level. It is quite possible that the supervision is "manufactured" by the programmer based on an application level information (as is common in many natural language applications [?]), or by another algorithm (e.g., an EM style algorithm or a semi-supervised learning algorithm). As mentioned earlier, we purposefully abstract away the learning algorithm itself and prefer to view the L operator essentially as the multiplication operation in conventional programming languages in that the specific details of the algorithm used are hidden from the programmer.

3.3.4. Inference

We have discussed so far the use of LBP to learn single variables. The capabilities provided by LBP at this level are already quite significant in that it significantly simplifies the generation of programs that interact with data and rely on a hierarchy or learning operators.

However, making decisions in real world problems often involve assigning values to *sets* of variables where a complex and expressive structure can influence, or even dictate, what assignments are possible. For example, in the task of labeling part-of-speech tags to the words of a sentence, the prediction is governed by the constraints like "no three consecutive words are verbs." Another example exists in scene interpretation tasks where predictions must respect constraints that could arise from the nature of the data or task specific conditions.

The study of how these joint predictions can be optimized globally via inference procedures, and how these may affect the training of the classifiers is an active research area. In LBP, the learning task is decoupled from the task of respecting the mutual constraints. Only after variables of interest are learned they used to produce global output consistent with the structural constraints.

LBP supports two general inference procedure a sequential conditional model (e.g., as in [17,15]) and a more general optimization procedure implemented via linear programming [18]. LBP is designed so that global inference for a collection of variables is an integral part of the language. The programmer needs to specify the variables involved which will be, in general, induced variables whose exact definition is determined in run time by LBP. The constraints among them can be either observed from the data (as in a sequential process) or specified as rules. In either case, an inference algorithm will be used to make global decision over the learned variables and constraints.

3.3.5. Compilation

An LBP program is a set of possible programs, where the specific element in the space of all programs is determined by a process of *data driven compilation*. This novel notion of data-driven compilation is fundamental to LBP. It combines optimization ideas from the theory of compilation with machine learning based components – RGFs defined in the program are becoming explicit functions as a result of the interaction of the program with data.

Developing a learning centered systems today requires an elaborate process of chaining classifiers. For example, in the context of natural langauge applications, extracting information from text requires first learning a part of speech (POS) tagger. Then, the available data is annotated using the learned POS tagger, and that data is used to (generate features and) learn how to decompose sentences into phrases. The new classifiers are then used to annotate data that is, in turn, used to learn tags required for the information extraction task. This sequential pipelining process has some conceptual problems, that have to do with error accumulation and avoiding inherent interactions across layers of computations. These can be dealt with algorithmically, and the inference capabilities provided by LBP will allow programmers to develop approaches to this problem.

Here, however, we would like to address the software engineering issue raised by this sequential process. Observing the computation tree required to extract the specific phrases of interest, for example, may reveal that most of the computation involved in this process is superfluous. Only a small number of the evaluations done are actually needed to support the final decisions. In traditional programming languages, compilation methods have been developed that can optimize the program and generate only the code that is actually required. LBP provides a framework within which this can be done also for learning intensive programs. We intend to develop optimization methods in the context of learning based programming that will perform analogous tasks and determine what needs to be learned and evaluated in the course of the data driven compilation.

3.4 Related Work

There are several research efforts that are related to the LBP research program. Perhaps the most related program, at least superficially, is "Yet Another Learning Environment" (YALE) developed at the University of Dortmund (http://yale.cs.uni-dortmund.de). Yale provides a unified framework for diverse machine learning applications. However, YALE is very different from LBP, both conceptually and functionally. It does not attempt to deal with raw data but rather with preprocessed data, it does not have feature extraction and example extraction mechanisms, and it does not deal with relational features as in LBP. It does not plan to support inference and does not have any view on compilation, and clearly not on data-driven compilation. Essentially, none of the issues presented in Sec. 3 are handled by YALE, which is a well designed package for integrative use of machine learning libraries, but not a programming paradigm. Other extensive machine learning libraries, like MLC++, are different for the same reasons. There are some other programs that have some conceptual similarity to LBP, but they typically represent more restricted efforts at developing programming languages for specific applications, such as constructing robot control software [20,12] or stochastic programs [14,16].

3.5 Discussion

The step represented by the Learning Based Programming paradigm is essential in order to develop large scale systems that acquire the bulk of their knowledge from raw, real world data and behave robustly when presented with new previously unseen situations.

This line of research can be viewed both as an end point of a line of research in machine learning – building on the maturity level the machine learning community has reached in understanding and in the design of classifiers - and as a beginning of a promising new programming discipline. At the same time, we expect that the availability of the LBP vehicle will motivate the study of new learning protocols as well as open up possibilities for experimental studies in intelligent interaction that are difficult to pursue today.

While the difficulties of developing machine learning centered systems are apparent to most researchers, there isn't yet a real research effort to develop an appropriate programming paradigm. In the preliminary development of LBP we attempt also to place this direction on the research agenda, given that developing a mature programming language is a huge effort that may require a community wide participation.

We believe that more research into fundamental issues from learning, compilation and software engineering perspectives, as well as a more mature implementation of LBP and several test cases of using it are required before we can determine the success of this approach as well as some of the key directions to follow up on.

3.6 Acknowledgments

This research is supported by NSF grants ITR-IIS-0085836, ITR-IIS-0085980 and IIS-9984168 and an ONR MURI Award.

References

1. S. Agarwal, A Awan, and D. Roth. Learning to detect objects in images via a sparse, part-based representation. *IEEE Transactions on Pattern Analysis and Machine Intelligence*, 20(11):1475–1490, 2004.
2. A. Blum. Learning boolean functions in an infinite attribute space. *Machine Learning*, 9(4):373–386, 1992.
3. A. Carlson, C. Cumby, J. Rosen, and D. Roth. The SNoW learning architecture. Technical Report UIUCDCS-R-99-2101, UIUC Computer Science Department, May 1999.
4. C. Cumby and D. Roth. Relational representations that facilitate learning. In *Proc. of the International Conference on the Principles of Knowledge Representation and Reasoning*, pages 425–434, 2000.
5. C. Cumby and D. Roth. Learning with feature description logics. In *Proc. Of the International Conference Inductive Logic Programming*, 2002.
6. C. Cumby and D. Roth. Feature extraction languages for propositionalized relational learning. In *IJCAI Workshop on Learning Statistical Models from Relational Data*, 2003.
7. Y. Even-Zohar and D. Roth. A classification approach to word prediction. In *Proc. Of the Annual Meeting of the North American Association of Computational Linguistics (NAACL)*, pages 124–131, 2000.
8. A. R. Golding and D. Roth. A Winnow based approach to context-sensitive spelling correction. *Machine Learning*, 34(1-3):107–130, 1999.
9. S. Har-Peled, D. Roth, and D. Zimak. Constraint classification for multiclass classification and ranking. In *The Conference on Advances in Neural Information Processing Systems (NIPS)*. MIT Press, 2002.
10. R. Khardon, D. Roth, and R. Servedio. Efficiency versus convergence of boolean kernels for on-line learning algorithms. In *The Conference on Advances in Neural Information Processing Systems (NIPS)*. MIT Press, 2001.
11. R. Khardon, D. Roth, and L. G. Valiant. Relational learning for NLP using linear threshold elements. In *Proc. of the International Joint Conference on Artificial Intelligence (IJCAI)*, pages 911–917, 1999.
12. H. J. Levesque, R. Reiter, Y. Lespérance, F. Lin, R. B. Scherl. Golog: a logic programming language for dynamic domains. *Journal of Logic Programming*, 31:59–84, 1997.
13. J. W. Lloyd. *Foundations of Logic Progamming*. Springer-Verlag, 1987.
14. D. McAllester, D. Koller, and A. Pfeffer. Effective bayesian inference for stochastic programs. In *Proceedings of the National Conference on Artificial Intelligence*, 1997.

15. A. McCallum, D. Freitag, and F. Pereira. Maximum entropy Markov models for information extraction and segmentation. In *Proceedings of ICML-00*, Stanford, CA, 2000.

16. A. Pfeffer. Ibal: An integrated bayesian agent language. In *Proceedings of the International Joint Conference of Artificial Intelligence*, 2001.

17. V. Punyakanok and D. Roth. The use of classifiers in sequential inference. In *The Conference on Advances in Neural Information Processing Systems (NIPS)*, pages 995–1001. MIT Press, 2001.

18. V. Punyakanok, D. Roth, W. Yih, and D. Zimak. Semantic role labeling via integer linear programming inference. In *Proc. the International Conference on Computational Linguistics (COLING)*, Geneva, Switzerland, August 2004.

19. D. Roth and W. Yih. Relational learning via propositional algorithms: An information extraction case study. In *Proc. of the International Joint Conference on Artificial Intelligence (IJCAI)*, pages 1257–1263, 2001.

20. S. Thrun. A framework for programming embedded systems: initial design and results. Technical Report CMU-CS-98-142, CMU Computer Science Department, 1998.

4 N-1 Experiments Suffice to Determine the Causal Relations Among N Variables

Frederick Eberhardt, Clark Glymour[1], Richard Scheines

Carnegie Mellon University

Abstract

By combining experimental interventions with search procedures for graphical causal models we show that under familiar assumptions, with perfect data, N - 1 experiments suffice to determine the causal relations among N>2 variables when each experiment randomizes at most one variable. We show the same bound holds for adaptive learners, but does not hold for N > 4 when each experiment can simultaneously randomize more than one variable. This bound provides a type of ideal for the measure of success of heuristic approaches in active learning methods of causal discovery, which currently use less informative measures.

4.1 Introduction

Consider situations in which the aim of inquiry is to determine the causal structure of a system with many variables, for example the gene regulation network of a species in a particular environment. The aim in other words is to determine for each pair of variables X, Y in a set of variables, S, whether X *directly* causes Y (or vice-versa), with respect to the remaining variables in S. We define direct causation as follows. If there is some set of values V for the variables besides X and Y such that X and Y covary in a population in which we randomly assign values to X while holding all other variables fixed at V, then X is a direct cause of Y. Such a system of causal relations can be represented by a directed graph, in which the variables are nodes or vertices of the graph, and X • Y indicates that X is a direct cause of Y. If there are no feedback relations among the variables, the graph is acyclic. For example, consider the three variables Exposed (to other children with active Chicken Pox), Infected (with the Chicken

---

[1] Second affiliation: Florida Institute for Human and Machine Cognition

F. Eberhardt et al.: *N-1 Experiments Suffice to Determine the Causal Relations Among N Variables*, StudFuzz **194**, 97–112 (2006)
www.springerlink.com

Pox virus), and Symptomatic (have the rash or not). Then the correct causal graph among these variables is as shown in Figure 1:

Figure 1: Causal Graph for Chicken Pox

Clearly there is no arrow from Exposure to Symptomatic, because no matter whether we fix Infection = yes or Infection = no, Exposure and Symptoms will not covary in a population in which we randomly assign Exposure.

In general, causal discovery concerns finding the most efficient way to determine the complete structure of such a directed acyclic graph, under some simplifying assumptions. The simplifying assumptions typically involve the *Causal Markov Assumption*, which says that every variable X is probabilistically independent of all of its non-effects, conditional on its immediate causes. In the example above that would entail that Symptomatic is independent of Exposure given Infection. It is also typical to assume the *Faithfulness Assumption*, which says that *all* of the conditional independence relations are consequences of the *Causal Markov Assumption* applied to the directed graph representing the causal relations. The Faithfulness Assumption guarantees, for example, that the two paths from Tax Rate to Tax Revenues (one negative through Economic Activity and one direct and positive) do not exactly balance thereby leaving Tax Rate and Tax Revenue independent, an independence that is not a consequence of the Causal Markov Assumption.

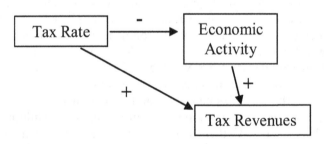

Figure 2: Tax Rate, Economic Activity and Tax Revenues

Even assuming the Causal Markov and Faithfulness conditions, causal discovery is difficult. Suppose that, before collecting data, nothing is known that will provide positive or negative evidence about the influence of any of the variables on any of the others. There are several ways to obtain data and to make inferences:

1. Conduct a study in which all variables are passively observed, and use the inferred associations or correlations among the variables to learn as much as possible about the causal relations among the variables.
2. Conduct an experiment in which one variable is assigned values randomly (randomized) and use the observed associations or correlations among the variables to learn as much as possible about the causal relations.
3. Do (2) while intervening to hold some other variable or variables constant.

Procedure 1. is characteristic of non-experimental social science, and it has also been proposed and pursued for discovering the structure of gene regulation networks (Spirtes, et. al, 2001). Consistent algorithms for causal inferences[2] from such data have been developed in computer science over the last 15 years. Some of these algorithms are based on conditional independence facts - the PC-Algorithm, for example (Spirtes, et al., 2000) - and others are based on assignments of prior probabilities and computation of posterior probabilities from the data (Meek, 1996; Chickering, 2002). We will appeal to facts about such procedures in what follows, but the details of the algorithms need not concern us.

There are, however, strong limitations on what can be learned from data that satisfy these assumptions, even supplemented with other, ideal simplifications, for example that there are no unrecorded common causes of the variables (we say the variable set is *causally sufficient*), and there are no feedback relations among the variables. Under these assumptions, and the assumption that we have the true joint probability distribution on the variables, then these algorithms can determine from the observed associations whether it is true that X and Y are adjacent, i.e., *whether* X *directly causes* Y *or* Y *directly causes* X, for all variables X, Y, but only in certain cases can the direction of causation be determined.

---

[2] By a consistent search algorithm we mean an algorithm that provably recovers as much about the graphical structure as is determined by the joint probability distribution, for all graphs and probability distributions satisfying the assumptions specified.

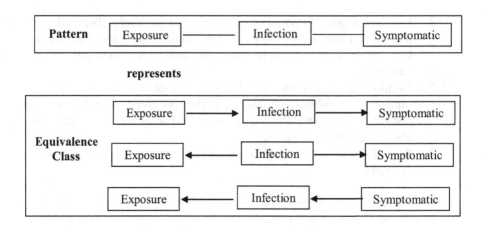

Figure 3: Equivalence Class

For example, in the Chicken Pox example, without any background knowledge or time order, the best a causal discovery algorithm could do from observational data on Exposure, Infection, and Symptoms is the equivalence graphs shown on the bottom of Figure 3, represented by the pattern of adjacencies on the top of Figure 3. No information about the direction of the arrows is inferable from just passively observed associations.

If, however, the true graph were the one we show in Figure 4, then a consistent causal discovery algorithm could identify all the relations just from passive observational data.

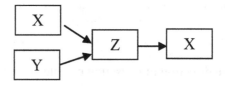

Figure 4

For several reasons, experimental intervention is a preferred method of estimating causal relations, and it is the standard method described in many methodology texts. If, for example, instead of passively observing the associations between Exposure, Infection, and Symptoms, we randomly assigned Exposure, then from the resulting associations we could uniquely determine the correct graph among these variables.

Randomization guards against the possibility that there is an unrecorded common cause of the manipulated variable and other variables. By randomizing X and observing which other variables in a set covary with X, we can determine which variables are influenced by X, but we cannot determine which variables influence X, as we have eliminated their influence by our random assignment of X's value. If, for example, in the Chicken Pox example we had chosen to perform an experiment in which we randomly assigned values for Symptoms, then we would have observed that Exposure and Infection are still associated, but that neither is associated with Symptoms. We would thus learn very little from this experiment about the true causal structure.

In even moderately complicated cases there is no single variable we can randomize that will tell us uniquely the true causal graph. For example, no single experiment in which we randomize one of either X or Y or Z uniquely distinguishes among the two structures in Figure 5:

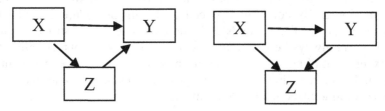

Figure 5

As it turns out, however, if we first perform an experiment in which we randomize one variable (it doesn't matter which), learn as much as we can, and then perform another in which we randomize either one of the remaining variables, then we can uniquely identify the causal graph among X, Y, and Z, no matter what it is. So a sequence of two experiments is sufficient to identify the graph, no matter what it is, among three variables. Using procedure 3, which involves randomly assigning values to one variable while holding other variables fixed doesn't help.

An *experiment* as we conceive it here, is an assignment of some variables to be passively observed and others to be randomly assigned their values. Thus, on three variables X, Y, and Z, there are 8 possible "experiments:

1. Passively Observe: {X, Y, Z}   Randomly Assign: { }
2. Passively Observe: {X, Y}   Randomly Assign: {Z}
3. Passively Observe: {X, Z}   Randomly Assign: {Y}
4. Passively Observe: {Y, Z}   Randomly Assign: {X}
5. Passively Observe: {X}   Randomly Assign: {Y, Z}
6. Passively Observe: {Z}   Randomly Assign: {X, Y}
7. Passively Observe: {Y}   Randomly Assign: {X, Z}
8. Passively Observe: { }   Randomly Assign: {X,Y,Z}

If we restrict our attention to only experiments that involve randomly assigning values to at most one variable, as described in procedures 1 and 2 above, then there are only 4 possible experiments among three variables.

In many domains, for example, genetics, scientists can and do perform sequences of experiments with the goal being to learn as much as possible about the causal structure of the variables they are manipulating and observing. It is nearly always time consuming and in many cases quite expensive to perform each experiment, however. Thus it would be useful to have a principled grip on 1) what can we learn from a given sequence of experiments of a certain sort, 2) how many experiments of a given sort are required, in the worst case, to learn the truth, and 3) what is the optimal sequence of experiments for causal discovery.

In this paper we propose a principled result that answers in a limited way question 2: By combining procedure 1 with procedure 2, under the assumptions so far listed, for $N > 2$, the complete causal structure on N variables can always be determined with $N - 1$ experiments, counting the null experiment of passive observation (procedure 1) as one experiment, if conducted. Further, this is the best possible result when at most one variable is randomized in each experiment.

4.2 The Idea

Consider the case of $N = 3$ variables. There are 25 directed acyclic graphs on 3 vertices. In Figure 6 we show the graphs sorted into sub-classes that are indistinguishable without experimental intervention. Given the joint distribution prior to an intervention, a consistent search algorithm will return the equivalence class of the graph - that is, the disjunction of all graphs in the box to which the true graph belongs.

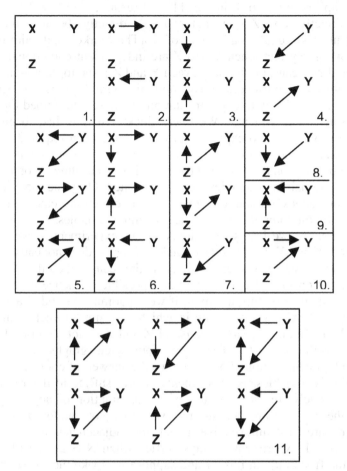

Figure 6: All graphs among X, Y, Z, broken into equivalence classes with respect to observational data

An experimental intervention, say one that randomizes X, provides extra information: the fact that the distribution of X has been randomly assigned by an external intervention tells us that in the resulting experimental data none of the remaining variables influence X. We can use that as prior information in applying a consistent search algorithm to the experimental data. If X is manipulated, the resulting joint distribution on X,Y and Z will give us, through such search procedures, information about whether X is a direct cause of Y or Z, or an indirect cause of one or the other. Thus suppose we randomize X and we find the following in the experimental distribution: Y and Z covary with X, and Y and Z are independent conditional on X. Then we know that X causes Y and Z, which tells us that the true graph is in box 6 or in box 11, and further, we know Y does not cause Z directly, and Z does not cause Y directly, because they are independent conditional on all values of X. (The Markov and Faithfulness assumptions imply that when Y and Z are independent conditional on X, there is no direct causal relation between Y and Z.) The top graph in box 6 must therefore be the true graph. By combining search procedures (in this case used informally) with experimentation, we have determined the truth with a single experiment. (We were lucky: if we had begun by randomizing Y or Z, two experiments would have been required.) When we randomize X and follow up with a consistent search procedure, which requires no additional experimentation, all of the direct connections between the remaining variables can be estimated. Only the directions of some of the edges remain unknown. Those directions can clearly be determined by randomizing each of the remaining variables.

In some cases, we lose something when we experiment. If when X is randomized, X and Y do not covary, we know that X does not cause Y, but we do not know whether Y causes X or neither causes the other, because our manipulation of X has destroyed any possible influence of Y on X. Thus in the single structure in box 9, if we randomize X, and Y and Z do not covary with X, every structure in which X is not a direct or indirect cause of Y or Z, and Y is not a cause of Z and Z is not a cause of Y, is consistent with our data. There are four such graphs. Subsequent experimental manipulation of Y will tell us the answer, of course.

So, with N variables, N experiments clearly suffice to determine the structure uniquely. In fact, because of the assumption of acyclicity and because the associations among the variables not randomized in an experiment are still informative about the adjacencies among these variables, N - 1 experiments always suffice when N > 2. To illustrate, consider the first graph in box 11 and suppose we make the worst choices for the sequence of variables to randomize: first X, then Y, then Z. Randomizing X we find only that X does not cause Y or Z, and that Y and Z are adjacent. Randomizing Y, we find that Y does cause X but does not cause Z, and that X and Z are adjacent. We reason as follows: Z must be a direct cause of X, because we know they are adjacent but we now know X is not a cause of Z.

Similarly, Z must be a direct cause of Y because they are adjacent and Y does not cause Z. Y must be a direct cause of X, because Y is a cause of X and Y is not a cause of Z (so there cannot be a pathway Y→Z→X). We have found the true graph, and only 2 experiments were required. We show in the appendix that the same result, that at most N-1 experiments are required, is true for all N > 2.

The result does not hold for N = 2, where there are only 3 possible structures: no edges, X→Y, and X←Y. Suppose X→Y. Suppose we randomize nothing, merely observe non-experimental values. If we find X, Y are associated, then a second experiment is required to determine the direction of the effect. Suppose instead, we begin by randomizing X. If we find X, Y are not associated, a second experiment is required to determine whether Y causes X.

The proof of the bound has three perhaps surprising corollaries. (1) Any procedure that includes passive observation in which no variables are randomized exceeds the lower bound for some cases, when the passive observation is counted as an experiment. (2) Controlling for variables by experimentally fixing their values is never an advantage. (3) "Adaptive" search procedures (Murphy, 1998; Tong and Koller, 2001) choose the most "informative" next experiment given the results of previous experiments. That is, they choose the next experiment that maximizes the expected information to be obtained. We also show that no adaptive procedure can do better than the N-1 lower bound on the number of experiments required to identify the structure in the worst case.

Implementations of adaptive search procedures generally have to make simplifying assumptions in order to make updating of the distribution over all possible graphs computationally tractable or must use greedy heuristics to select the next intervention that is deemed most informative with respect to the underlying causal structure given the evidence from the previous experiments. The success of such a heuristic is commonly measured by comparing it to a strategy that randomly chooses the next experiment no matter what the evidence is so far. Other comparisons are to uniform sampling or to passive observation - the latter obviously only provides limited directional information. These comparisons indicate whether the heuristic is achieving anything at all but give little insight into how well the strategy compares with an ideal procedure. The bound we provide specifies such an ideal, at least in the case when only passive observational and single intervention experiments are considered.

4.3 Discussion

A variety of theoretical issues remain. Expected complexities can differ considerably from worst case complexities, and we have not investigated the expected number of experiments required for various probability distributions on graphs. When the variable set is not known to be causally sufficient, which is the typical scientific situation, there is a consistent search procedure, the FCI Algorithm (Spirtes et al., 1993; 2000), which unsurprisingly returns more limited information about the structure. When there are unrecorded common causes, some structures cannot be distinguished by independence and conditional independence relations alone, but can be distinguished by attention to changes in the covariation of variables in different experiments. In general, the number of experiments required is larger than N-1 but we have no bound to report. Further, we do not know by how much the N - 1 bound can be improved by experiments in which two or more variables are randomized simultaneously. However, for N>4, multiple intervention experiments do reduce the total number of experiments required to identify the causal structure even in the worst case, although it may initially seem that information on the potential edges between intervened upon vertices is lost. For example, in the case of five vertices, three such multiple simultaneous randomization experiments suffice even in the worst case; the same is true for six vertices: three experiments will do.

The N-1 bound can be considered proportional to a minimum cost of inquiry when all experiments, including passive observation, have the same cost. Costs may differ when one experiment simultaneously randomizes several variables. In practice there is a cost to sample size as well, which can result in complicated trade-offs between cost and the confidence one has in the results. In practice, with real data, search procedures tend to be unreliable for dense graphs. The reasons differ for different algorithms, but basically reflect the fact that in such graphs conditional probabilities must be assessed based on many conditioning variables. Each conditioning set of values corresponds to a subsample of the data, and the more variables conditioned on, the smaller the sample, and the less reliable the estimate of the conditional joint probability of two variables. Some search algorithms, such as PC and FCI, test for conditional independence relations and use the results as an oracle for graphical specification. So it would be of interest to know the effects on the worst case number of experiments required when a bound is placed on the number of variables conditioned on in the search procedure.

Acknowledgements

We are grateful to Teddy Seidenfeld, who contributed to early discussion on the work in this paper and asked several key questions that helped to focus our efforts.

The second author is supported by ONR grant N00014-04-1-0384 and NASA grants NCC2-1399 and NCC 2-1377. The third author is supported by a grant from the James S. McDonnell Foundation.

Appendix: Proofs

We assume the reader's familiarity with some fundamental ideas from directed graphical models, or Bayes nets, including the property of d-separation (Pearl, 1988) and search algorithms that exploit that property (Spirtes, et. al, 2000).

Assumptions:

We make the following assumptions in our proof of the worst case bound on the number of experiments required to identify the causal graph underlying N variables.

Faithfulness: The distribution over the variables is faithful to a directed acyclic graph on the variables in the data.

Full scale D-Separation: It is possible to condition on any subset of variables to determine d-separation relations.

Perfect Data: The data is not supposed to be of any concern. In particular, we are not concerned with weak causal links, insufficient or missing data. The data is such that we can identify the conditional independencies if there are any.

Interventions: Interventions are possible on every variable.

Definitions:

An **experiment** randomizes at most one variable and returns the joint distribution of all variables.

A **procedure** is a sequence of experiments and a structure learning algorithm applied to the results of these experiments.

A procedure is **reliable** for an N vertex problem if and only if for all DAGs on N vertices the procedure determines the correct graph uniquely.

A procedure is **order reliable** for an N vertex problem if and only if it is reliable for all non-redundant orderings of experiments.

A procedure is **adaptive** if and only if it chooses at each step one from among the possible subsequent experiments as a non-trivial function of the results of the previous experiments.

Claims

Proposition 1: *For N > 2, there is an order reliable procedure that in the worst case requires no more than N - 1 experiments, allowing only single interventions.*

Proof: Consider a graph with N vertices where N > 2 and let $X_1,..., X_N$ specify an arbitrary ordering of these vertices. Let each experiment consist of an intervention on one variable. Perform N-1 experiments, one intervention on each X_i where $1 \leq i \leq$ N-1. By Lemma 1 below, applying the PC-algorithm to the first experiment determines the adjacencies among at least $X_2,..., X_N$. The k-th experiment determines the directions of all edges adjacent to X_k: if X_j is adjacent to X_k, then X_k is a direct cause of X_j if and only if X_j covaries with X_k when X_k is randomized (since if X_k were only an indirect cause of X_j, and since X_j and X_k are adjacent, X_j would have to be a direct cause of X_k, and there would be a cycle); otherwise, X_j is a direct cause of X_k. X_N has not been randomized, but its adjacencies with every other variable have been determined by the N-1 experiments. Suppose X_N and X_k are adjacent. Since X_k has been randomized, X_k is a cause of X_N if and only if X_N covaries with X_k when X_k is randomized. In that case, if X_k were an indirect but not a direct cause of X_N, then X_N would be a direct cause of X_k, because X_N and X_k are adjacent, and hence there would be a cycle. If X_N and X_k do not covary when X_k is randomized, then, since they are adjacent, X_N is a direct cause of X_k. If X_k and X_N are not adjacent, then this missing edge would have been identified in one of the interventions on X_j, where $j \neq k$. These are all of the cases. Q.E.D.

Lemma 1: *If G is a causal graph over a set of variables V, and G' the manipulated graph resulting from an ideal intervention on variable X in G,* *then for all pairs of variables Z, Y distinct from X, Z and Y are d-separated by some S ⊆ V in G if and only if Z and Y are d-separated by some S' ⊆ V in G'.*

Proof: G' is identical to G except that all edges into X in G do not occur in G'.

L-to-R: First assume Z and Y are d-separated by some S ⊆ V in G. Then no undirected path between Z and Y in G d-connects those variables relative to S. Suppose for reductio that Z and Y are not d-separated by S in G'. Then some path between Z and Y in G' must now be active, i.e., constitutes a d-connection. The paths between Z and Y in G' are a subset of those in G. Thus some path between Z and Y that was inactive in G must now be active in G'. Thus all nodes on such a path that were inactive in G must now be active in G'. But if X was inactive on a path in G relative to S, it will still be inactive in G' relative to S. For it to be otherwise, X would either have to switch from a non-collider to a collider, which cannot happen by removing edges into X, or for X to be a collider in G with no ancestor in S but to be a collider in G' with an ancestor, which also cannot happen by removing edges into X. A similar argument applies equally to non-X nodes, so Z and Y are d-separated by S in G'.

R-to-L: Next assume that Z and Y are not d-separated by some S ⊆ V in G, that is, they are d-connected by every S ⊆ V in G. Then Z and Y are adjacent in G, and an intervention on X does not remove this adjacency, thus they are still adjacent in G' and thus d-connected by every S ⊆ V in G'. Q.E.D.

Proposition 2: *No order reliable procedure randomizing a single variable at each step requires fewer than N -1 experiments for an N variable problem in the worst case.*

Proof: In order to show that N-1 experiments are in the worst case necessary given N variables let X_1, ..., X_N again specify an arbitrary ordering of the N vertices. Suppose only N-2 interventions were performed in sequence, one each on X_1 to X_{N-2}. Suppose that in the true underlying causal graph X_{N-1} and X_N happen to both be (direct) causes of each X_i, where $1 \leq i \leq$ N-2, and that X_{N-1} and X_N are adjacent. It does not matter in which direction this edge is pointing, but assume, without loss of generality, that X_N is a parent of X_{N-1}. Note that in this case all of the interventions on X_1, ..., X_{N-2} will indicate that there is an edge between X_N

and X_{N-1}, but none will be able to direct it. Hence, an (N-1)th experiment is required. Q.E.D.

Comment: A similar situation occurs when each X_i, where $1 \leq i \leq N-2$, is a (direct) common cause of X_N and X_{N-1} and when, again, X_N is the parent of X_{N-1} or vice versa. Here also, none of the N-2 experiments will be able to identify the direction of the edge between X_N and X_{N-1}. It follows that N-1 experiments are sufficient and in the worst case necessary to identify the causal graph underlying N vertices. N - 1 is a tight bound for the worst case number of single intervention experiments.

The fact that the sequence of experimental interventions is arbitrary in the previous proof suggests that this result is still true for the worst case even when the choice of the next experiment is adaptive, that is, even if at each point during the sequence of experiments the "best" experiment given the evidence from the previous experiment is chosen. Although Proposition 3 follows from the previous two proofs as a corollary, the proof below emphasizes the aspect that no *adaptive* strategy will do any better in the worst case.

Proposition 3: *Every reliable adaptive procedure for which each experiment randomizes a single variable requires, in the worst case, at least N -1 experiments for an N vertex problem.*

Proof: Clearly N-1 experiments are sufficient to identify the causal graph underlying N vertices since they are sufficient for the non-adaptive case. In the following we will show that in the worst case N - 1 experiments are necessary even if an adaptive strategy is adopted for the experimental sequence. The situation can be viewed as a game between experimenter and nature: The experimenter specifies an experiment and nature returns the independence relations true of the graph, possibly modified by the experimental intervention. At each point in the game, however, nature may return the independence relations implied by the largest equivalence class of graphs that are consistent with the independence relations supplied to the experimenter in the previous experiments. The claim of proposition 3 amounts then to the claim that there always exists a strategy for nature that ensures that the experimenter requires N-1 experiments to reduce the equivalence class of graphs over the N variables to one, i.e. to identify the underlying causal structure uniquely. Consequently, no matter what the experimenter's adaptive strategy is, it will take N-1 experiments in this game.

Nature's strategy is as follows: Let V_1, V_2,... be the sequence of variables the experimenter intervenes upon. When the experimenter intervenes upon variable V_i, nature maintains the equivalence class of graphs that satisfy the following conditions: The class contains *all* the graphs that have complete subgraphs among the non-intervened variables, i.e. $V_{i+1},...,V_N$ and V_k is a direct cause of V_j, where $1 \leq k \leq i$ and $1 \leq j \leq k$ with $j \neq k$. In other words, whichever variable the experimenter intervenes upon, it is the sink of all the variables that have not yet been intervened upon, that is V_N is a (direct) parent of all other vertices, V_{N-1} is a parent of $V_1,...,$ V_{N-2} etc. Then V_2 is a parent of V_1 only and V_1 is a child of all other vertices. Note, that the equivalence class of graphs nature maintains is the set of graphs which are all isomorphic to each other among the variables not intervened upon.

Now consider the adaptive strategy of the experimenter trying to identify the graph. At each stage in the game, she has no information about the directions of the edges among the non-intervened variables. So, in particular, after N-2 experiments, she has no information on the direction of the edge between V_{N-1} and V_N. Hence an (N-1)th experiment is required. It follows that even with an adaptive strategy, N-1 experiments are in the worst case necessary to identify the causal graph among N variables. Q.E.D.

Other Types of Experiments

In the previous two proofs an experiment was always assumed to consist of an intervention on one particular variable. However, it might be thought that other types of experiments, such as passive observations or interventions on more than one variable might improve the worst case result of N-1 experiments. While it is true that multiple interventions (randomizing more than one variable at a time) can shorten the experimental sequence, this is not the case for passive observational studies. We call a passive observational experiment a null-experiment.

The above proofs indicate that the worst case always occurs for particular complete graphs. If one were to run a null-experiment at any point in the experiment sequence when the underlying graph is complete - the most likely time would probably be at the beginning - then one would realize that one is confronted with a complete graph. However, this information (and more) is obtained anyway from two sequential experiments, each consisting of an intervention on a particular variable. The null-experiment paired with any other experiment cannot generate more information about the graph than two single intervention experiments, since a single intervention experiment also identifies all adjacencies except for those into the intervened variable. But a second intervention on a different variable would identify these interventions, too. So the only advantage of the null-experiment is in the case where only one experiment is run. The above proofs only apply to graphs of three or more variables, which certainly cannot always be identified by one experiment alone. In fact, even for two variables, two experiments are needed in the worst case (see discussion in main body of the paper).

References

1. D.M. Chickering (2002). Learning Equivalence Classes of Bayesian-Network Structures. *Journal of Machine Learning Research*, 2:445-498.
2. C. Meek, (1996). PhD Thesis, Department of Philosophy, Carnegie Mellon University
3. K.P. Murphy, (2001). Active Learning of Causal Bayes Net Structure, Technical Report, Department of Computer Science, U.C. Berkeley.
4. J. Pearl, (1988). *Probabilistic Reasoning in Intelligent Systems*, San Mateo, CA. Morgan Kaufmann.
5. J. Pearl, (2000). *Causality*, Oxford University Press.
6. P. Spirtes, C. Glymour and R. Scheines, (1993). *Causation, Prediction and Search*, Springer Lecture Notes in Statistics, 1993; 2nd edition, MIT Press, 2000.
7. P. Spirtes, C. Glymour, R. Scheines, S. Kauffman, W. Aimalie, and F. Wimberly, (2001). Constructing Bayesian Network Models of Gene Expression Networks from Microarray Data, in *Proceedings of the Atlantic Symposium on Computational Biology, Genome Information Systems and Technology*, Duke University, March.
8. S. Tong and D. Koller, (2001). Active Learning for Structure in Bayesian Networks, *Proceedings of the International Joint Conference on Artificial Intelligence*.

5 Support Vector Inductive Logic Programming

S.H. Muggleton[1], H. Lodhi[2], A. Amini[2] and M.J.E. Sternberg[2]

1. Department of Biological Sciences, Imperial College London
2. Department of Computing, Imperial College London

Abstract

In this paper we explore a topic which is at the intersection of two areas of Machine Learning: namely Support Vector Machines (SVMs) and Inductive Logic Programming (ILP). We propose a general method for constructing kernels for Support Vector Inductive Logic Programming (SVILP). The kernel not only captures the semantic and syntactic relational information contained in the data but also provides the flexibility of using arbitrary forms of structured and non-structured data coded in a relational way. While specialised kernels have been developed for strings, trees and graphs our approach uses declarative background knowledge to provide the learning bias. The use of explicitly encoded background knowledge distinguishes SVILP from existing relational kernels which in ILP-terms work purely at the atomic generalisation level. The SVILP approach is a form of generalisation relative to background knowledge, though the final combining function for the ILP-learned clauses is an SVM rather than a logical conjunction. We evaluate SVILP empirically against related approaches, including an industry-standard toxin predictor called TOPKAT. Evaluation is conducted on a new broad-ranging toxicity dataset (DSSTox). The experimental results demonstrate that our approach significantly outperforms all other approaches in the study.

5.1 Introduction

In this paper we propose a novel machine learning approach which combines the dimensionality independence advantages of Support Vector Machines (SVMs) with the expressive power and flexibility of Inductive

S.H. Muggleton et al.: *Support Vector Inductive Logic Programming*, StudFuzz **194**, 113–135 (2006)
www.springerlink.com   © Springer-Verlag Berlin Heidelberg 2006

Logic Programming (ILP). In particular, we propose a kernel which is an inner product in the feature space spanned by a given set of first order clauses. As with normal ILP, examples, background knowledge and hypothesised clauses are encoded as logic programs. The kernel not only captures the semantic and syntactic relational information contained in the data but also provides the flexibility of using arbitrary forms of structured and non-structured data.

The approach we suggest differs from the relational kernels suggested in [10,11] by our use of logical background knowledge. In order to understand the distinction being made here consider the following three settings for ILP.

Atomic generalisation.   This setting is characterised by having examples which are typically ground atomic formulae and hypotheses consisting of atomic formulae which entail the examples. Reynolds [40] and Plotkin [37] showed that this hypothesis space forms a lattice which is partially ordered by atomic subsumption.

Clausal generalisation.   In this setting examples are ground clauses and hypotheses are clauses which entail the examples. Plotkin [36,38] showed that once more this hypothesis space forms a lattice which is partially ordered by clausal subsumption.

Clausal generalisation relative to background knowledge.   This third setting [38] is distinguished by assuming the existence of background knowledge in the form of a conjunction of clauses. Examples are ground clauses. Hypotheses are clauses which when conjoined with the background knowledge entail the examples.

Most ILP research has assumed the third setting, clausal generalisation relative to background knowledge, since this is the more general approach. The use of background knowledge provides a flexible way of encoding the understanding of domain experts, and can increase both the predictive accuracy of the learning and the degree of insight provided relative to the background knowledge. However, this setting brings with it overheads related to the theorem proving involved in using background knowledge. For this reason Page and Frisch [34] investigated the use of atomic generalisation with respect to a monadic constraint theory. This is a generalisation of the first setting, and a special case of the third setting.

More recently Lloyd [22] and others [11] have investigated algorithms which use the setting of atomic generalisation, but with more general forms of strongly-typed terms. In particular, terms can consist of arbitrary sets. This allows more flexibility for defining data types without the overheads associated with background knowledge. In [11] it is shown that this form of representation and learning can be used to formulate a relational kernel. In [10] it is shown that by using the "bag of atoms" representation introduced in [2] a multi-instance kernel approach can even be applied to structurally complex ILP learning problems involving small molecules.

The SVILP approach is a form of generalisation relative to background knowledge, though the final combining function for the ILP-learned clauses is an SVM rather than a logical conjunction.

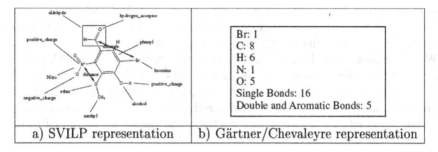

| a) SVILP representation | b) Gärtner/Chevaleyre representation |

Fig. 1. Molecule represented using a) SVILP representation which employs a kernel based on domain-expert informed chemical background knowledge indicated by the annotations on the figure and b) Gärtner/Chevaleyre bag-of-atoms uses Multi-Instance (MI) kernel based on frequency of occurrences of atoms and atom pairs.

We will now provide a simplified worked example to show the difference in representing molecules using the Gärtner/Chevaleyre approach from the representation used by our SVILP kernel. Figure 1 shows a typical molecule from the DSSTox dataset of toxins (see Section 5.1). In the SVILP approach we start by formulating chemical background knowledge in the form of Prolog definitions. These have been designed by one of the authors (Ata Amini), a biochemistry domain expert, to be relevant to properties associated with toxins. Such properties include the existence of substructures such as aromatic rings, methyl and alcohol substructures, types of atom, charge, the existence of hydrogen acceptors and distances between various critical structures on the molecule. The ILP system Progol is used to generate a set of hypothesised clauses based on the given background knowledge and examples. An SVM kernel is then used as the combining function for predictions of these clauses. By contrast, the Gärtner/Chevaleyre features consist simply of the frequency of occurrence of atoms, bonds and atom pairs within the given molecule.

These are used to form a vector representation of the molecule. An obvious advantage of this "bag-of-atoms" representation is that it requires no domain expertise and thus is less effort to develop. By analogy with the use of the "bag-of-words" [1,8] representation in text classification one might expect a simple representation of this form to lead to superior predictive accuracy. However, this is not the case in the experiments reported in Section 5.6 in which the SVILP kernel significantly outperforms the Gärtner/Chevaleyre kernel. In this case, the use of more highly informed background knowledge in the SVILP appears to provide a significant advantage.

The chapter is arranged as follows. The Background Section 5.2 introduces the basic ideas behind kernels, SVMs and Inductive Logic Programming (ILP). In Section 5.3 SVILPs are defined and their properties proved. This is followed by a section which describes Related Work (Section 5.4). In Section 5.5 the implementation of our SVILP is described together with an analysis of its complexity. Next we describe the Experiments (Section 5.6) on toxicity data, which demonstrate that SVILP produces low predictive error compared with alternative feature-based kernels, and when combined with such kernels predictive error is further reduced. The paper then concludes.

5.2 Background

5.2.1. Kernels and Support Vector Machines

During recent years, there has been increasing interest in kernel-based methods such as Support Vector Machines (SVMs) [48,6]. The non-dependence of these methods on the dimensionality of the feature space and the flexibility of using any kernel make them a good choice for different tasks such as classification and regression.

Let S be a set of n training instances of the form

$$S = \{(d_1, c_1), (d_2, c_2), \cdots\cdots, (d_n, c_n)\}$$

Here d_i are the instances, found in the instance space D (not necessarily a vector space, any finite set) and $c_i = c(d_i)$ are the labels or response variables. For binary classification $c_i \in \{+1, -1\}$ otherwise $c_i \in \{1, 2, ..., k\}$. Note that for regression $c_i \in R$. We can view the learning process of SVMs as comprising two stages.

1. Map the input data into some higher dimensional space H through a non-linear mapping ϕ that is given by

$$\phi : D \to H.$$

The mapping ϕ may not be known explicitly but be accessed via the kernel function described below.

2. Construct a linear function f in the space.

The kernel function $K(d_i, d_j) = \langle \phi(d_i), \phi(d_j) \rangle$ computes the inner product between the mapped instances.

Support Vector Classification: The support vector machine for classification (SVC) is based on the idea of constructing the maximal margin hyperplane in feature space. This unique hyperplane separates that data into two categories $\{-1, +1\}$ with maximum margin and is given by

$$f(d) = \langle w, \phi(d) \rangle + b.$$

Here w is the weight vector learned during the training phase. This weight vector is a linear combination of training instances. In other words $w = \sum_{i=1}^{n} \alpha_i \phi(d_i)$. To find a maximal margin hyperplane one has to solve a convex quadratic optimisation problem. The corresponding classification function is given by

$$f(d) = sgn(\sum_{i=1}^{n} \alpha_i c_i K(d_i, d) + b).$$

Hence the classifier is constructed only using the inner products between the mapped instances.

Support Vector Regression: SVMs used for regression (SVR) inherit many of the properties that characterise SVMs for classification. SVR embeds the input data into Hilbert space through a non-linear mapping ϕ and constructs a linear regression function in this space. In order to apply the support vector technique to regression tasks a reasonable loss function is used. An ε-insensitive loss function is a popular choice that is defined by $|c - f(d)|_\varepsilon = \max(0, |c - f(d)| - \varepsilon)$. The loss function allows error below some $\varepsilon > 0$ and controls the width of the insensitive band. Regression estimation is performed solving an optimisation problem. The regression function f is given by

$$f(d) = \sum_{i=1}^{n} (\alpha_i^\star - \alpha_i) K(d_i, d) + b.$$

where α_i and $\alpha_i^\star$ are Lagrange multipliers.

We now briefly describe a kernel function that is the building block of these methods. As described earlier this is a function that calculates the inner product between mapped examples. The kernel computes this inner product by implicitly mapping the examples to the space. The mathematical foundation of such a function was established during the first decade of the twentieth century [25]. A kernel function is a symmetric function

$$K(d_i, d_j) = K(d_j, d_i) \quad \text{for } i, j = 1, ..., n,$$

and satisfies the following property of positive (semi)definiteness.

$$\sum_{1, j=1}^{n} a_i a_j K(d_i, d_j) \geq 0 \quad \text{for } a_i, a_j \in R.$$

The $n \times n$ matrix with entries of the form $K_{ij} = K(d_i, d_j)$ is known as the kernel matrix or the Gram matrix. A kernel matrix is a symmetric, positive definite matrix. In other words the n Eigen values of this $n \times n$ kernel matrix are non-negative.

Linear, polynomial and Gaussian Radial Basis Function (RBF) kernels are well-known examples of general purpose kernel functions. A linear kernel function is given by

$$K_{linear}(d_i, d_j) = K(d_i, d_j) = d_i' d_j.$$

Given a kernel K the polynomial construction is given by

$$K_{poly}(d_i, d_j) = (K(d_i, d_j) + r)^p,$$

where p is a positive integer and r is a nonnegative constant. Clearly, we incur a small computational cost, to define a new feature space. The feature space corresponding to a degree p polynomial kernel includes all products of at most p input features. Hence, the polynomial kernel create images of the examples in a feature space having a huge number of dimensions. Note that for $p = 1$ we get the linear construction.

Furthermore, the RBF kernel defines a feature space with an infinite number of dimensions. Given a set of vectors the Gaussian RBF kernel is given by

$$K_{RBF}(d_i, d_j) = \exp(\frac{-\|d_i - d_j\|^2}{2\sigma^2}).$$

The RBF kernel allows an algorithm to learn a linear function in a feature space having an infinite number of dimensions.

Kernel functions are closed under addition and multiplication. This property enables the design of new compound kernels. We illustrate this idea by a simple example. Let K_1 be a kernel that computes the inner product between the mapped molecules in a feature space generated by numeric chemical features. Let the kernel K_2 calculate the inner product between the molecules that are mapped into a space generated by relational features. A kernel that sums the respective inner products will be a new kernel. Formally

$$K = K_1 + K_2$$

Similarly

$$K = K_1 * K_2$$

$$K = aK_1$$

Kernel functions can be defined over general sets [43,49,12]. This important fact has allowed successful exploration of novel kernels for discrete spaces such as strings, graphs and trees [23,3,17]. String kernels computes an inner product in the feature space generated by all subsequences of a given length. The kernel is based on the idea of comparing two strings by means of the subsequences they contain. The more subsequences two strings have in common, the more similar they are considered. An important part is that such substrings do not need to be contiguous, and the degree of contiguity of one such subsequence in a string determines how much weight it will have in the comparison.

5.2.2 Inductive Logic Programming

Inductive Logic Programming (ILP) [26,32,29] is the area of AI which deals with the induction of hypothesised predicate definitions. In ILP logic programs are used as a single representation for examples, background knowledge and hypotheses. ILP is differentiated from most other forms of Machine Learning (ML) both by its use of an expressive representation language and its ability to make use of logically encoded background knowledge. This has allowed successful applications of ILP [30] in areas such as molecular biology [47,5,16] and chemoinformatics [9,46].

In the following it is assumed that the examples, background knowledge and hypotheses each consist of logic programs, ie sets of first-order Horn clauses. The normal semantics of ILP is as follows. We are given background (prior) knowledge B and evidence E. The evidence $E = E^+ \wedge E^-$ consists of positive evidence E^+ and negative evidence E^-. The aim is then to find a hypothesis H such that the following conditions hold.

Prior Satisfiability. $B \wedge E^- \nvDash \Box$

Posterior Satisfiability. $B \wedge H \wedge E^- \nvDash \Box$

Prior Necessity. $B \nvDash E^+$

Posterior Sufficiency. $B \wedge H \vDash E^+$

Since a large number of hypotheses will typically fit such a definition, the Bayesian ILP setting [27] assumes a prior probability distribution defined over the hypothesis space. Algorithms such as CProgol [28] use such a prior to search for hypotheses which maximise the posterior probability $p(H \mid E)$.

5.3 Support Vector Inductive Logic Programming (SVILP)

The SVILP framework builds on the ILP framework. Thus we also assume background knowledge B, examples E and a hypothesis H for which the conditions of the normal semantics hold. The key difference between ILP and SVILP is the way in which the set of clauses H is used for predictive purposes. In ILP H is simply treated as a conjunction, for which any instance d from the domain of instances D is predicted to be true if and only if

$$B, H \vDash d$$

By contrast, SVILP bases a kernel on the predictions of the clauses h in H. This involves forming a binary hypothesis-instance association matrix M in which element $M_{ij} = 1$ (0 otherwise) if and only if clause $h_i \in H$ entails instance $d_j \in D$ as follows.

$$B, h_i \vDash d_j$$

The kernel described in Section 5.2 can be viewed as a function for which similarity of two instances d_1 and d_2 is based on the similarity of the rows of clauses in M associated with d_1 and d_2.

5.3.1 Family example

In this artificial example we assume that the occurrence of a disease is related to the inheritance patterns of an observable property (eg. hair colour) in various families. The background knowledge is shown in Figure 2. This describes the relationships in the family tree shown in Figure 3. Examples of individuals having the disease are shown in Figure 4 and various hypothesised clauses are shown in Figure 5. Assuming the domain is limited to the examples, we show the resulting binary hypothesis-instance association matrix in Figure 6.

```
father(henry,john).        father(david,henry).    mother(jane,john).
mother(elizabeth,henry).   father(charles,mary).   father(egbert,jill).
mother(jill,mary).         mother(ann,jill).

grandfather(F,P) ← father(F,P1), parent(P1,P).
grandmother(M,P) ← mother(M,P1), parent(P1,P).

parent(F,P) ← father(F,P).
parent(M,P) ← mother(M,P).

hair(john,blond).       hair(mary,black).       hair(jane,black).
hair(henry,blond).      hair(charles,black).    hair(david,black).
hair(jill,blond).       hair(elizabeth,blond).  hair(ann,blond).
hair(egbert,black).
```

Fig. 2. Background knowledge for disease inheritance

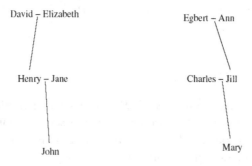

Fig. 3. Family trees for disease inheritance

1. disease(john)
2. disease(mary)
3. disease(jane)
4. disease(henry)
5. disease(charles)

Fig. 4. Examples for disease inheritance

A. disease(P)   ← hair(P,Colour), father(F,P), hair(F,Colour)
B. disease(P)   ← hair(P,Colour), mother(M,P), hair(M,Colour)
C. disease(P)   ← hair(P,Colour),grandmother(M,P), hair(M,Colour)
D. disease(P)   ← hair(P,Colour), grandfather(F,P), hair(F,Colour)
E. disease(P)   ← hair(P,black), father(F,P), mother(M,P),
                  hair(F,blond), hair(M,black)
F. disease(P)   ← hair(P,black), father(F,P), grandfather(G,P),
                  hair(F,blond), hair(G,black)

Fig. 5. Hypothesised clauses for disease inheritance

	A	B	C	D	E	F
1	1	0	0	1	0	0
2	1	0	0	1	0	0
3	1	0	0	0	0	0
4	0	1	0	0	0	0
5	1	0	0	0	0	0

Fig. 6. Resulting binary hypothesis-instance association matrix

Note that according to the matrix examples 1 and 2 have maximum similarity. This is despite the fact that the hair colour (the main observable feature) of John and Mary (the individuals involved in the examples) are opposite (blond and black respectively). The example demonstrates the strong learning bias which can be introduced by the use of background knowledge and hypotheses within the SVILP setting. In the next section we define the kernel formally.

5.3.2 Definition of kernel

We assume background knowledge B and a set of hypothesised clauses H drawn from a class of hypotheses H and a set of instances D drawn from a class of instances D. Each hypothesis clause h in H can be thought of as a function of the following form.

$$h : D \to \{\text{True}, \text{False}\}$$

Conversely the τ function gives the hypothesised clauses covering any particular instance.

$$\tau : D \to 2^H$$

where for any d_i in D

$$\tau(d_i) = \{h : \exists h \in H, (B, h \models d_i)\}$$

As in the Bayesian ILP framework [27], we assume a prior probability distribution over the hypotheses. This can be represented as a function π such that

$$\pi : H \to [0,1]$$

and

$$\sum_{h \in H} \pi(h) = 1$$

Next we define a function, which maps sets of hypothesised clauses to probabilities.

$$f : 2^H \to [0,1]$$

For all $H' \subseteq H$

$$f(H') = \sum_{h \in H'} \pi(h)$$

Now the kernel function is as follows.

$$K : D \times D \to [0,1]$$

For all d_i, d_j in D

$$K(d_i, d_j) = f(\tau(d_i) \cap \tau(d_j))$$

It can be easily shown that the kernel is an inner product in ILP space. The kernel requires a hypothesised clause set H. In order to improve the informative power of the kernel we define a prior probability distribution and fits the prior to the coordinates in space spanned by the hypothesised clauses. In this way a countable set of hypothesised clauses implies a mapping ϕ that maps the data into an ILP space, where dimensionality of the space is the same as the cardinality of the set of hypothesised clauses. Formally

$$f_i(d) = \sqrt{\pi_i(h_i(d))} \quad \text{for } i = 1, \dots, k$$

Hence the mapping ϕ for an instance is given by

$$\phi : d \to ((f_1(d), f_2(d), \dots, f_k(d)) = (f_i(d)_{i=1}^k)$$

and kernel for instance d_i and d_j is given by

$$K(d_i,d_j) = \langle \phi(d_i),\phi(d_j) \rangle$$
$$= \sum_{i=1}^{k} f_i(d_i)f_i(d_j)$$

Hence

$$K(d_i,d_j) = f(\tau(d_i) \cap \tau(d_j))$$

The validity of kernel function follows from the definition as an inner product however we can show that it satisfies Mercer's condition (symmetry and positive definiteness). Clearly the kernel function is symmetric and positive definiteness occurs since there is mapping ϕ from D into an ILP space. For all $a_i \in R$ and $d_i \in D$, for $i = 1,...,n$ we have the following expression,

$$\sum_{1,j=1}^{n} a_i a_j K(d_i,d_j)$$

We now use a compact representation $A = (a_i)_{i=1}^{n}$ and $\phi = (\phi(d_i))_{i=1}^{n}$, hence kernel matrix $\sum_{i,j=1}^{n} K(d_i,d_j) = \phi\phi'$

$$= A\phi'\phi A'$$
$$= t't \geq 0$$

Given that ϕ maps the data into ILP space, we can construct Gaussian RBF kernels in ILP space

$$K_{RBF}(d_i,d_j) = \exp(\frac{-\|\phi(d_i)-\phi(d_j)\|^2}{2\sigma^2}).$$

where

$$\|(\phi(d_i)-\phi(d_j)\| = \sqrt{K(d_i,d_i)-2K(d_i,d_j)+K(d_j,d_j)}.$$

Our method is flexible to construct any kernel in the space spanned by the clauses. However we select RBF kernels, (K_{RBF}), constructed in ILP space for our experiments in section 5.6

5.4 Related work

5.4.1 Propositionalisation

Within ILP "propositionalisation" [19] techniques transform machine learning problems from a first-order logic setting into one which can be handled by a "propositional" or feature-based learner. Kramer et al. [19] distinguish between domain-independent ([21,20,44,7]) and domain-dependent approaches (eg [18]). In most domain-independent propositionalisation approaches [20,44,7] features are introduced as clauses with a monadic head predicate. For instance, when applied to problems involving molecular descriptions these techniques introduce new features such as the following.

```
f1(A)  :- has_rings(A, [R1, R2]),
          hydrophobic(A,H),
          H > 1.0.
F2(A)  :- atm(A,B_,27, _),
          Bond(A,B,C, _),
          Atm (A,C, _,29, _).
```

Though superficially similar to domain-independent propositionalisation, the SVILP approach described in this paper is not a propositionalisation technique since it does not transform the representation by the introduction of such monadic features. Instead a general-purpose ILP learning algorithm is used to learn clauses with heads having arbitrary predicate arities. The heads of these clauses can contain terms with multi-arity function symbols and constants. In normal ILP the hypothesis used for predictive purposes would consist of these clauses conjoined together. In SVILP the truth-value predictions of these individual clauses are projected onto the instance space. The kernel matrix is then formulated over the instance-space predictions of the individual clauses.

SVILP is similar in its use of support-vector technology to the domain-dependent propositionalisation approach of Kramer and Frank [18]. This uses bottom-up evaluation to find features representing high-frequency graph fragments from a molecular database. The key difference here is that SVILP is domain-independent, allowing the use of background knowledge to encode the appropriate machine learning bias.

5.4.2 Kernels within ILP

Within ILP there has recently been interest in the development of kernels which incorporate relational information, for use within support vector machines. Several authors [11,24] take the approach of using syntactic measures of distance between first-order formulae as the basis for such kernels. Within the ILP literature it is normal to differentiate between *syntactic* [14,39,13] and *semantic* [33] distance measures. Syntactic measures are based on differences in the structure of first-order formulae, and tend to be confined to comparison of terms, rather than arbitrary first-order formulae. Semantic measures are based on comparison of models, making this approach intractable for all but simple formulae.

The kernel approaches described in [11,24] are unable to make use of background knowledge, since they are based on syntactic comparison of ground atoms. By contrast, a central feature of the SVILP described in this paper is its use of generalisation relative to background knowledge.

5.5 Implementation

Algorithm Construct Kernel

Given Background Knowledge B, Examples E, Instance domain D

Call CProgol5.0 on B, E to produce H

Comp t the hypothesis-instance matrix M
Foreach d_i in D
Foreach h_j in H
If ($B, h_j \models d_i$) **Then** $M_{ij} = 1$
Else $M_{ij} = 0$

Compute the Kernel K using M
Foreach d_i in D
Foreach d_j in D

$$K(d_i, d_j) = f(\tau(d_i) \cap \tau(d_j))$$
Return K

Fig. 7. Algorithm for computing the SVILP kernel

Figure 7 shows the algorithm for computing the kernel. Examples and Background knowledge are given to CProgol5.0 [28,31] which generates a set of hypothesised clauses with associated probabilities. The hypothesised clauses are then taken by a Prolog program which computes the hypothesis-instance association M (see Section 5.3), indicating for each instance the set of all hypothesised clauses which imply it. Implication is tested using proof-length bounds which are the same as those provided by the user for the CProgol5.0 run. Next the kernel matrix K is computed for each pair of instance by taking intersections and unions over the sets in M.

Assuming the theorem prover can test each hypothesised clause against each instance in time bounded by a constant k, the overall time taken to compute the kernel is proportional to the number of hypothesised clauses $|H|$ and the number of instances $|D|$.

5.6 Experiments

A new dataset was used for evaluating SVILP. The DSSTox dataset was made available to us by Dr Ann Richards of National Health and Environmental Effects Research Laboratory, USA. The dataset represents the most diverse set of toxins presently available in the public domain. By choosing a new toxin dataset we avoided over-testing problems associated with molecular datasets such as the Mutagens [15,45]. The 188 molecule Mutagenic dataset has now been evaluated by so many researchers that it is becoming hard to argue that some of the higher reported accuracies are not simply due to chance.

5.6.1 Materials

DSSTox

The DSSTox [41,42] database contains organic and organometalic molecules with their toxicity values. The dataset consists of 576 molecules. Figure 8 shows an example of two of the molecules found in DSSTox. As far as we know no previous attempt has been made to quantify the structure and activity relationship for the whole DSSTox dataset.

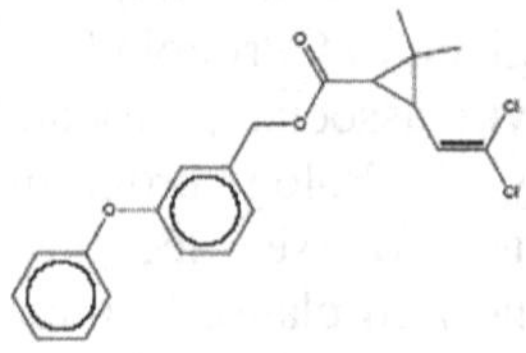

Permethrin

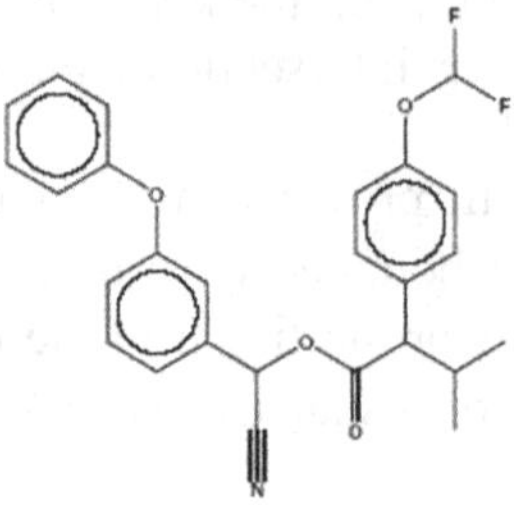

Flucythrinate

Fig. 8. Examples of compounds in DSSTox

5.6.2 Methods

We now describe the pre-processing stage. Molecules in the form of SMILES strings, were transformed into 3D structures using the software CONCORD 4.0 [35] (implemented in TRIPOS).

All of the molecules contain continuous chemical feature known as LUMO, dipole moment and LOGP. In order to define LOGP, we briefly describe the organic phase and aqueous phase. A compound in a mixture of an organic molecule and water is distributed between two phases, namely the organic phase and the aqueous phase. LOGP is given by $LOGP = log(\frac{C_{organic}}{C_{aqueous}})$, where $C_{organic}$ is the concentration in the organic phase and $C_{aqueous}$ is the concentration in aqueous phase. Molecules having large LOGP are hydrophobic molecules whereas hydrophilic molecules have small LOGP. LUMO (Lowest Unoccupied Molecular Orbital) describes the electrophilicity of the molecules. LOGP describes the passage of molecules through the membrane based on their hydrophobicity, whereas LUMO and similar electronic properties indicate the ability of compounds to react covalently with biological molecules.

The key information is given in the form of atom and bond description. Consider a simple example:

Example:

```
Aldehyde(M,A!) :- atom(M,A1,c,2,-,-,-,-),
atom(M,A2,o,2,-,-,-,-),
atom(M,A3,h,-,-,-,-,-),
bond(M,A1,A2,2),
bond(M,A1,A3,1).
```

LOGP	LUMO	Positive
Charge	Negative	Charge
COO	NH2	C=C
CC	CN	Hydrogen Acceptor
Hydrogen Donor	NO2	S-S
COOH	Amide	Ether
Thioether	Epoxy	Hetero-nonaromatic-6ring
Hetero-nonaromatic-5ring	Hetero-aromatic-6ring	Hetero-aromatic-5ring
Phenyl	Cyclohexane	Cyclopentane
Naphthalene	Pyridine	Methyl
Ethyl	Propyl	Pentyl
Hexyl	Bigalkyl	Isopropyl
Isobutyl	Tert-butyl	Fluorine
Chlorine	Bromine	Iodine
Phosphorous	Sulphur	Aromatic
Nitrogen	SP3 Nitrogen	SP2 Nitrogen
SP Nitrogen	SP2 Oxygen	SP3 Oxygen
aldehyde	dipole moment	

Fig. 9. The full list of chemical features in the background knowledge

We compared the performance of SVILP with a number of related techniques including partial least squares (PLS), multi instance kernels (MIK) [10,11], an RBF kernel using only 3 chemical features (LOGP, LUMO, dipole moment) that we term as CHEM. We also compared the performance of SVILP with well known QSAR software TOPKAT (Toxicity Prediction by Komputer Assisted Technology).

As our experimental methodology we used 5-fold cross validation. For evaluation we used mean squared error (MSE) and R-squared (standard measure of accuracy in QSAR). In this work we employed ε-insensitive SVM regression (SVR)[48]. We used the SVM package SVMTorch [4] for our experiments.

C, ε, σ are the tunable parameters for kernel-based methods (SVILP, CHEM and MIK). In PLS the tunable parameter is the "number of components". These parameters can be set by some model selection method. The traditional protocol to set the values for the parameters is the minimisation (maximisation) of some criterion relative to the values of the parameters using a validation set. We select the optimal values of the tunable parameters using a validation set as described. We set the parameters for each fold using only the training set of the fold. We randomly selected a subset comprising 75% of the data (training set of each fold) for the training set and used the remaining data as a test set. A range of values of the parameters were selected. The sets of the values are given by $C = \{10, 100, 1000, 10000\}$, $\varepsilon = \{0.1, 0.3, 0.5, 1.0\}$, $\sigma = \{0.125, 0.25, 0.5, 4, 16\}$. For PLS we used the number of components from 1 to 15. The parameters which give the minimum MSE on the validation set were chosen.

In order to perform the prediction task using SVILP, we first obtained a set of clauses hypothesised by CProgol5.0. For all the folds, the clauses with positive compression were selected where the number of obtained clauses for each fold can vary between 1500-200. The compression value of a clause is given by

$$V = \frac{P * (p - (n + c + h))}{p},$$

where p is the number of positive instances correctly deducible from the clause, n is the number of negative examples incorrectly deducible from the clause, c is the length of the clause and h is number of further atoms to complete the clause and P is the total number of positive examples. In this work we used a uniform probability distribution over the clauses. Once we have a set of weighted clauses hypothesised by CProgol5.0, we computed the similarity between molecules using proposed kernel.

Results

We conducted an extensive series of experiments to evaluate the performance of the proposed method.

	MSE	R-squared
CHEM	0.811 ± 0.084	0.519 ± 0.031
PLS	0.671 ± 0.053	0.593 ± 0.016
MIK	0.838 ± 0.044	0.503 ± 0.034
SVILP	$\mathbf{0.574 \pm 0.051}$	$\mathbf{0.655 \pm 0.015}$

Fig. 10. MSE and R-squared with standard error for CHEM, PLS, MIK and SVILP.

	Accuracy
ILP	55
(PROGOL)	
CHEM	58
PLS	71
MIK	60
SVILP	**73**

Fig. 11. Accuracy for ILP, CHEM, PLS, MIK and SVILP.

	MSE	R-squared
CHEM	1.04	0.48
PLS	1.03	0.47
TOPKAT	2.2	0.26
SVILP	**0.8**	**0.57**

Fig. 12. MSE and R-squared for CHEM, PLS, TOPKAT and SVILP.

We conducted the first set of experiments to evaluate the efficacy of the new method for predicting the toxicity values. Figure 10 shows the results. The results are averaged over 5 runs of the methods. The results show that SVILP outperforms all the other methods in the study. We also evaluated our approach by employing it for categorising the molecules into two categories, toxic and non-toxic. We compared the performance of SVILP with the standard ILP system PROGOL. Figure 11 shows the results for the category "toxic". The result validate the efficacy of SVILP. In order to compare the performance of SVILP with TOPKAT we used a test set comprising 165 molecules. Figure 12 shows the results.

5.7 Conclusions and further work

In this paper we introduce a new framework for combining Support Vector machine technology with Inductive Logic Programming. Unlike existing relational kernels, the present approach works within the standard ILP setting of generalisation with respect to background knowledge, rather than the limited setting of atomic generalisation. A particular kernel is defined and implemented on top of the ILP system CProgol5.0. This kernel has been tested on an important new toxin dataset. In our experiments we compared the performance of the SVILP against both standard ILP and other relational and non-relational kernels. In all cases our kernel produced higher predictive accuracy.

Further theoretical work is necessary to clarify the effects on performance of varying the amount of background knowledge used by the kernel. The implementation of the system also needs to be better integrated. One way to do so might be to incorporate the kernel into a standard ILP system such as Progol. Also further empirical work is needed to test the kernel on a wider variety of relational problems.

Acknowledgements

The authors would like to acknowledge the support of the DTI Beacon project "Metalog - Integrated Machine Learning of Metabolic Networks Applied to Predictive Toxicology", Grant Reference QCBB/C/012/00003 and the ESPRIT IST project "Application of Probabilistic Inductive Logic Programming II (APRIL II)", GrantRef: FP-508861.

References

1. C. Apt, F.J. Damerau, and S.M. Weiss. "Automated learning of decision rules for text categorization." *ACM Trans on Information Systems*, *12*:pp.233-251, 1994.
2. Y.Chevaleyre and J.D. Zucker. "A framework for learning rules from multiple instance data. " Proceedings of the European Conference on Machine Learning (ECML 2001), pp.49--60, Berlin, 2001. Springer-Verlag. LNAI 2167.
3. M. Collins and N.Duffy. "Convolution kernels for natural language." *in Advances in Neural Information Processing System 14*. MIT Press, 2002.
4. R.Collobert and S.Bengio. "Svmtorch: Support vector machines for large-scale regression problems." *Journal of Machine Learning Research, 1*:pp.143-160, 2001.
5. A.Cootes, S.H. Muggleton, and M.J.E. Sternberg. "The automatic discovery of structural principles describing protein fold space." *Journal of Molecular Biology*, 2003.
6. N.Cristianini and J.Shawe-Taylor. "zAn introduction to Support Vector Machines." Cambridge University Press, Cambridge, UK, 2000.
7. L.Dehaspe and H.Toivonen. "Discovery of frequent datalog patterns." *Data Mining and Knowledge Discovery, 3(1)*:pp.7-36, 1999.
8. S.T. Dumais, J.Platt, D.Heckermann, and M.Sahami. "Inductive learning algorithms and representations for text categorisation." *in Proceedings of CIKM-98, 7th ACM International Conference on Information and Knowledge Management*, pp.148-155, 1998.

9. P.Finn, S.H. Muggleton, D.Page, and A.Srinivasan. "Pharmacophore discovery using the Inductive Logic Programming system Progol." *Machine Learning, 30*:pp.241-271, 1998.

10. Thomas Gartner, PeterA. Flach, Adam Kowalczyk, and AlexJ. Smola. "Multi-instance kernels." *in Proceedings of the Nineteenth International Conference on Machine Learning*, pp.176-186. Morgan-Kaufmann, 2002.

11. Thomas Gartner, JohnW. Lloyd, and PeterA. Flach. Kernels for structured data. In Stan Matwin and Claude Sammut, editors, Proceedings of the Twelfth International Conference on Inductive Logic Programming, LNAI 2583, pages 66--83, Berlin, 2002. Springer-Verlag.

12. 12.D.Haussler. "Convolution kernels on discrete structures."Technical Report UCSC-CRL-99-10, University of California in Santa Cruz, Computer Science Department, July 1999.

13. T.Horvath, S.Wrobel, and U.Bohnebeck. "Relational instance-based learning with lists and terms." Machine Learning, 43(1/2):53--80, 2001.

14. A.Hutchinson. "Metrics on terms and clauses." In M.Someren and G.Widmer, editors, Proceedings of the Ninth European Conference on Machine Learning, pages 138--145, Berlin, 1997. Springer.

15. R.D. King, S.H. Muggleton, A.Srinivasan, and M.Sternberg. "Structure-activity relationships derived by machine learning: the use of atoms and their bond connectives to predict mutagenicity by inductive logic programming." Proceedings of the National Academy of Sciences, 93:438--442, 1996.

16. R.D. King, K.E. Whelan, F.M. Jones, P.K.G. Reiser, C.H. Bryant, S.H. Muggleton, D.B. Kell, and S.G. Oliver. "Functional genomic hypothesis generation and experimentation by a robot scientist." Nature, 427:247--252, 2004.

17. R.I. Kondor and J.Lafferty. "Diffusion kernels on graphs and other discrete input spaces." In ICML, 2002.

18. S.Kramer and Frank E. "Bottom-up propositionalisation." In Proceedings of the ILP-2000 Work-In-Progress Track, pages 156--162. Imperial College, London, 2000.

19. S.Kramer, N.Lavrac, and P.Flach. "Propositionalisation approaches to Relational Data Mining." In S.Dzeroski and N.Larac, editors, Relational Data Mining, pages 262--291. Springer, Berlin, 2001.

20. S.Kramer, B.Pfahringer, and C.Helma. "Stochastic propositionalisation of non-determinate background knowledge." In Proceedings of the Eighth International Conference on Inductive Logic Programming, pages 80--94, Berlin, 1998. Springer-Verlag.

21. N.Lavra\vc, S.D\vzeroski, and M.Grobelnik. "Learning non-recursive definitions of relations with LINUS." In Yves Kodratoff, editor, Proceedings of the 5th European Working Session on Learning, volume 482 of Lecture Notes in Artificial Intelligence. Springer-Verlag, 1991.

22. J.W. Lloyd. "Logic for Learning." Springer, Berlin, 2003.

23. H.Lodhi, C.Saunders, J.Shawe-Taylor, N.Cristianini, and C.Watkins. "Text classification using string kernels." Journal of Machine Learning Research, 2002, (to appear).

24. D. Mavroeidis and P.A. Flach. "Improved distances for structured data." In T. Horvath and A. Yamamoto, editors, Proceedings of the Thirteenth International Conference on Inductive Logic Programming, LNAI 2835, pages 251--268, Berlin, 2003. Springer-Verlag.

25. J.Mercer. "Functions of positive and negative type and their connection with the theory of integral equations." Philosophical Transactions of the Royal Society London (A), 209:415--446, 1909.

26. S.H. Muggleton. "Inductive Logic Programming." New Generation Computing, 8(4):295--318, 1991.

27. S.H. Muggleton. "Bayesian Inductive Logic Programming." In M.Warmuth, editor, Proceedings of the Seventh Annual ACM. Conference on Computational Learning Theory, pages 3--11, New York, 1994. ACM Press. Keynote presentation.

28. S.H. Muggleton. "Inverse entailment and Progol." New Generation Computing, 13:245--286, 1995.

29. S.H. Muggleton. "Inductive logic programming: issues, results and the LLL challenge." Artificial Intelligence, 114(1--2):283--296, December 1999.

30. S.H. Muggleton. "Scientific knowledge discovery using Inductive Logic Programming." Communications of the ACM, 42(11):42--46, November 1999.

31. S.H. Muggleton and C.H. Bryant. "Theory completion using inverse entailment." In Proc.of the 10th International Workshop on Inductive Logic Programming (ILP-00), pages 130--146, Berlin, 2000. Springer-Verlag.

32. S.H. Muggleton and L.De Raedt. "Inductive logic programming: Theory and methods." Journal of Logic Programming, 19,20:629--679, 1994.

33. S.H. Nienhuys-Cheng. "Distance between Herbrand interpretations: a measure for approximations to a target concept." In N.Lavravc and S.Dvzeroski, editors, Proceedings of the Seventh International Workshop on Inductive Logic Programming (ILP97), pages 321--226, Berlin, 1997. Springer-Verlag. LNAI 1297.

34. D. Page and A. Frisch. "Generalization and learnability: A study of constrained atoms." In S.Muggleton, editor, Inductive Logic Programming. Academic Press, London, 1992.

35. R.S. Pearlman. "Concord User's Manual." Tripos, Inc, St Louis, Missouri, 2000.

36. G.Plotkin. "A further note on inductive generalization." In Machine Intelligence, volume6. Edinburgh University Press, 1971.

37. G.D. Plotkin. "A note on inductive generalisation." In B.Meltzer and D.Michie, editors, Machine Intelligence 5, pages 153--163. Edinburgh University Press, Edinburgh, 1969.

38. G.D. Plotkin. "Automatic Methods of Inductive Inference." PhD thesis, Edinburgh University, August 1971.

39. J. Ramon and M. Bruynooghe. "A framework for defining distances between first-order logic objects." In D. Page, editor, Proceedings of the Eighth International Workshop on Inductive Logic Programming (ILP98), pages 271-280, Berlin, 1998. Springer-Verlag. LNAI 1446.

40. J.C. Reynolds. "Transformational systems and the algebraic structure of atomic formulas. " In B.Meltzer and D.Michie, editors, Machine Intelligence 5, pages 135--151. Edinburgh University Press, Edinburgh, 1969.

41. A.M. Richard and C.R. Williams. "Distributed structure-searchable toxicity (DSSTox) public database network: A proposal." Mutation Research, 499:27--52, 2000.

42. C.L. Russom, S.P. Bradbury, S.J. Brodrius, D.E. Hammermeister, and R.A. Drummond. "Predicting modes of toxic action from chemical structure: Acute toxicity in the fathead minnow (oimephales promelas)." Environmental Toxicology and Chemistry, 16(5):948--967, 1997.

43. B.Scholkopf. "Support Vector Learning." R. Oldenbourg Verlag, Munchen, 1997. Doktorarbeit, Technische Universitat Berlin.Available from http://www.kyb.tuebingen.mpg.desimbs.

44. A.Srinivasan and R.King. "Feature construction with inductive logic programming: a study of quantitative predictions of biological activity aided by structural attributes." *Data Mining and Knowledge Discovery, 3(1)*:35--57, 1999.

45. A.Srinivasan, S.H. Muggleton, R.King, and M.Sternberg. "Theories for mutagenicity: a study of first-order and feature based induction." *Artificial Intelligence, 85(1,2)*:277--299, 1996.

46. M.J.E. Sternberg and S.H. Muggleton. "Structure activity relationships (SAR) and pharmacophore discovery using inductive logic programming (ILP)." *QSAR and Combinatorial Science, 22*, 2003.

47. M.Turcotte, S.H. Muggleton, and M.J.E. Sternberg. "Automated discovery of structural signatures of protein fold and function." *Journal of Molecular Biology, 306*:591--605, 2001.

48. V.Vapnik. "The Nature of Statistical Learning Theory." Springer Verlag, New York, 1995.

49. C.Watkins. "Dynamic alignment kernels." In A.J. Smola, P.L. Bartlett, B.Sch\"olkopf, and D.Schuurmans, editors, Advances in Large Margin Classifiers, pages 39--50, Cambridge, MA, 2000. MIT Press.

6 Neural Probabilistic Language Models

Yoshua Bengio[1], Holger Schwenk[2],
Jean-Sébastien Senécal[1], Fréderic Morin[1] and Jean-Luc Gauvain[2]

1. Département d'Informatique et Recherche Opérationnelle,
   Université de Montréal, Montréal, Québec, Canada
   bengioy@iro.umontreal.ca

2. Groupe Traitement du Langage Parlé
   LIMSI-CNRS, Orsay, France
   schwenk@limsi.fr

Abstract

A central goal of statistical language modeling is to learn the joint probability function of sequences of words in a language. This is intrinsically difficult because of the **curse of dimensionality**: a word sequence on which the model will be tested is likely to be different from all the word sequences seen during training. Traditional but very successful approaches based on n-grams obtain generalization by concatenating very short overlapping sequences seen in the training set. We propose to fight the curse of dimensionality by **learning a distributed representation for words** which allows each training sentence to inform the model about an exponential number of semantically neighboring sentences. Generalization is obtained because a sequence of words that has never been seen before gets high probability if it is made of words that are similar (in the sense of having a nearby representation) to words forming an already seen sentence. Training such large models (with millions of parameters) within a reasonable time is itself a significant challenge. We report on several methods to speed-up both training and probability computation, as well as comparative experiments to evaluate the improvements brought by these techniques. We finally describe the incorporation of this new language model into a state-of-the-art speech recognizer of conversational speech.

Y. Bengio et al.: *Neural Probabilistic Language Models*, StudFuzz **194**, 137–186 (2006)
www.springerlink.com

6.1 Introduction

A fundamental problem that makes language modeling and other learning problems difficult is the *curse of dimensionality*. It is particularly obvious in the case when one wants to model the joint distribution between many discrete random variables (such as words in a sentence, or discrete attributes in a data-mining task). For example, if one wants to model the joint distribution of 10 consecutive words in a natural language with a vocabulary V of size 100,000, there are potentially $100\,000^{10} - 1 = 10^{50} - 1$ free parameters. When modeling continuous variables, we obtain generalization more easily (e.g. with smooth classes of functions like multi-layer neural networks or Gaussian mixture models) because the function to be learned can be expected to have some local smoothness properties. For discrete spaces, the generalization structure is not as obvious: any change of these discrete variables may have a drastic impact on the value of the function to be estimated, and when the number of values that each discrete variable can take is large, most observed objects are almost maximally far from each other in Hamming distance.

A useful way to visualize how different learning algorithms generalize, inspired from the view of non-parametric density estimation, is to think of how probability mass that is initially concentrated on the training points (e.g., training sentences) is distributed in a larger volume, usually in some form of neighborhood around the training points. In high dimensions, it is crucial to distribute probability mass where it matters rather than uniformly in all directions around each training point.

A statistical model of language can be represented by the conditional probability of the next word given all the previous ones, since

$$\hat{P}(w_1^T) = \prod_{t=1}^{T} \hat{P}(w_t \mid w_1^{t-1}),$$

where w_t is the t-th word, and writing sub-sequence $w_i^j = (w_i, w_{i+1}, ..., w_{j-1}, w_j)$. Such statistical language models have already been found useful in many technological applications involving natural language, such as speech recognition, language translation, and information retrieval. Improvements in statistical language models could thus have a significant impact on such applications.

When building statistical models of natural language, one considerably reduces the difficulty of this modeling problem by taking advantage of word order, and the fact that temporally closer words in the word sequence are statistically more dependent. Thus, *n-gram* models construct tables of conditional probabilities for the next word, for each one of a large number of *contexts*, i.e. combinations of the last $n-1$ words:

$$\hat{P}(w_t \mid w_1^{t-1}) \approx \hat{P}(w_t \mid w_{t-n+1}^{t-1}).$$

We only consider those combinations of successive words that actually occur in the training corpus, or that occur frequently enough. What happens when a new combination of n words appears that was not seen in the training corpus? We do not want to assign zero probability to such cases, because new combinations are likely to occur, and they will occur even more frequently for larger context sizes. A simple answer is to look at the probability predicted using a smaller context size, as done in back-off trigram models [Katz, 1987] or in smoothed (or interpolated) trigram models [Jelinek and Mercer, 1980]. A way to understand how such models obtain generalization to new sequences of words is to think about a corresponding generative model. Essentially, a new sequence of words is generated by "gluing" very short and overlapping pieces of length 1, 2 ... or up to n words that have been seen frequently in the training data. The rules for obtaining the probability of the next piece are implicit in the particulars of the back-off or interpolated n-gram algorithm. Typically researchers have used $n = 4$, i.e. fourgrams, and obtained state-of-the-art results, but see [Goodman, 2001a] for how combining many tricks can yield to substantial improvements.

However, there is obviously much more information in the sequence that immediately precedes the word to predict than just the identity of the previous couple of words. In addition, this approach does not take advantage of a notion of similarity between words that would go beyond equality of words. For example, having seen the sentence "The cat is walking in the bedroom" in the training corpus should help us generalize to make the sentence "A dog was running in a room" almost as likely, simply because "dog" and "cat" (resp. "the" and "a", "room" and "bedroom", etc...) have similar semantic and grammatical roles.

There are many approaches that have been proposed to address these two issues, and we will briefly explain in Section 6.1.2 the relations between the approach proposed here and some of these earlier approaches. We will first discuss what is the basic idea of the proposed approach. A more formal presentation will follow in Section 6.2, using an implementation of these ideas that relies on shared-parameter multi-layer neural networks. Another very important part of this chapter are methods for efficiently training such very large neural networks (with millions of parameters) for very large data sets (with tens of millions of examples).

Many operations in this paper are in matrix notation, with lower case v denoting a column vector and v' its transpose, A_j the j-th row of a matrix A, and $x \cdot y = x'y$.

6.1.1 Fighting the Curse of Dimensionality with Distributed Representations

In a nutshell, the idea of the proposed approach can be summarized as follows:

1. associate with each word in the vocabulary a distributed *word feature vector* (a real-valued vector in $\mathbb{R}^m$),
2. express the joint *probability function* of word sequences in terms of the feature vectors of these words in the sequence, and
3. learn simultaneously the *word feature vectors* and the parameters of that *probability function*.

The feature vector represents different aspects of the word: each word is associated with a point in a vector space. The number of features (e.g. $m = 30...100$ in the experiments) is much smaller than the size of the vocabulary (e.g. 10 000 to 100 000). The probability function is expressed as a product of conditional probabilities of the next word given the previous ones, (e.g. using a multi-layer neural network to predict the next word given the previous ones, in the experiments). This function has parameters that can be iteratively tuned in order to **maximize the log-likelihood of the training data** or a regularized criterion, e.g. by adding a weight decay penalty[1]. The feature vectors associated with each word are learned, but they could be initialized using prior knowledge of semantic features.

Why does it work? In the previous example, if we knew that dog and cat played similar roles (semantically and syntactically), and similarly for (the,a), (bedroom,room), (is,was), (running,walking), we could naturally generalize (i.e. transfer probability mass) from

The cat is walking in the bedroom

to   A dog was running in a room

---

[1]Like in ridge regression, the squared norm of the parameters is penalized.

and likewise to

The cat is running in a room
A dog is walking in a bedroom
The dog was walking in the room

...

and many other combinations. In the proposed model, it will so generalize because "similar" words are expected to have a similar feature vector. Since the probability function is a *smooth* function of these feature values, a small change in the features will induce a small change in the probability. Therefore, the presence of only one of the above sentences in the training data will increase the probability, not only of that sentence, but also of its combinatorial number of "neighbors" in sentence space (as represented by sequences of feature vectors).

6.1.2 Relation to Previous Work

The idea of using neural networks to model high-dimensional discrete distributions has already been found useful to learn the joint probability of $Z_1 \cdots Z_n$, a set of random variables where each is possibly of a different nature [Bengio and Bengio, 2000a,b]. In that model, the joint probability is decomposed as a product of conditional probabilities

$$\hat{P}(Z_1 = z_1, \cdots, Z_n = z_n)$$
$$\overset{def}{=} \prod_i \hat{P}(Z_i = z_i \mid g_i(Z_{i-1} = z_{i-1}, Z_{i-2} = z_{i-2}, \cdots, Z_1 = z_1)),$$

where $g(\cdot)$ is a function represented by a neural network with a special left-to-right architecture, with the i-th output block $g_i()$ computing parameters for expressing the conditional distribution of Z_i given the value of the previous Z's, in some arbitrary order. Experiments on four UCI data sets showed this approach to work comparatively very well [Bengio and Bengio, 2000a,b]. Here, however, we must deal with data of variable length, like sentences, so the above approach must be adapted. Another important difference is that here, all the Z_i (word at i-th position) refer to the same type of object (a word). The model proposed here therefore introduces a sharing of parameters across time – the same g_i is used across time – and across input words at different positions. It is a successful large-scale application of the idea in [Bengio and Bengio,

2000a,b], along with the older idea of learning a distributed representation for symbolic data, that was advocated in the early days of connectionism [Hinton, 1986; Elman, 1990]. More recently, Hinton's approach was improved and successfully demonstrated on learning several symbolic relations [Paccanaro and Hinton, 2000]. The idea of using neural networks for language modeling is not new either, e.g. [Miikkulainen and Dyer, 1991]. In contrast, here we push this idea to a **large scale**, and concentrate on learning a **statistical model** of the distribution of word sequences, rather than learning the role of words in a sentence. The approach proposed here is also related to previous proposals of character-based text compression using neural networks to predict the probability of the next character [Schmidhuber, 1996]. The idea of using a neural network for language modeling has also been independently proposed by [Xu and Rudnicky, 2000], although experiments are with networks without hidden units and a single input word, which limits the model to essentially capturing unigram and bigram statistics.

The idea of discovering some similarities between words to obtain generalization from training sequences to new sequences is not new. For example, it is exploited in approaches that are based on learning a clustering of the words [Brown et al., 1992; Pereira et al., 1993; Niesler et al., 1998; Baker and McCallum, 1998]: each word is associated deterministically or probabilistically with a discrete class, and words in the same class are similar in some respect. In the model proposed here, instead of characterizing the similarity with a discrete random or deterministic variable (which corresponds to a soft or hard partition of the set of words), we use a continuous real-vector for each word, i.e. a **learned distributed feature vector**, to represent similarity between words. The experimental comparisons in this paper include results obtained with class-based n-grams [Brown et al., 1992; Ney and Kneser, 1993; Niesler et al., 1998], which are based on replacing the sequence of past words representing the prediction context by the corresponding sequence of word classes (the class of a word being the identity of its cluster).

The idea of using a vector-space representation for words has been well exploited in the area of *information retrieval*, for example see work by [Schutze, 1993], where feature vectors for words are learned on the basis of their probability of co-occurring in the same documents (see also Latent Semantic Indexing [Deerwester et al., 1990]). An important difference is that here we look for a representation for words that is helpful in representing compactly the probability distribution of word sequences from natural language text. Experiments suggest that learning jointly the representation (word features) and the model is very useful. The idea of using a continuous representation for words has been exploited successfully by [Bellegarda, 1997] in the context of an n-gram based statistical language model, using LSI to dynamically identify the topic of discourse. Finally, the approach discussed here is close in spirit to the more recent research on discovering an *embedding* for words or symbols, in a low-dimensional space, as in [Hinton and Roweis, 2003] and [Blitzer *et al.*, 2005].

The idea of a vector-space representation for symbols in the context of neural networks has also previously been framed in terms of a parameter sharing layer, for protein secondary structure prediction [Riis and Krogh, 1996], and text-to-speech mapping [Jensen and Riis, 2000)].

6.2 A Neural Model

The training set is a sequence $w_1 \cdots w_T$ of words $w_t \in V$, where the vocabulary V is a large but finite set. The objective is to learn a good model $f(w_t, \cdots, w_{t-n+1}) = \hat{P}(w_t \mid w_{t-n+1}^{t-1})$, in the sense that it gives high out-of-sample likelihood. In the experiments, we will report the geometric average of $1/\hat{P}(w_t \mid w_{t-n+1}^{t-1})$, also known as *perplexity*, which is also the exponential of the average negative log-likelihood. The only constraint on the model is that for any choice of w_{t-n+1}^{t-1}, $\sum_{i=1}^{|V|} f(i, w_{t-1}, \cdots, w_{t-n+1}) = 1$, with $f > 0$. By the product of these conditional probabilities, one obtains a model of the joint probability of sequences of words.

We decompose the function $f(w_t, \cdots, w_{t-n+1}) = \hat{P}(w_t \mid w_{t-n+1}^{t-1})$ in two parts:

1. A mapping C from any element i of V to a real vector $C(i) \in \mathbb{R}^m$. It represents the *distributed feature vectors* associated with each word in the vocabulary. In practice, C is represented by a $|V| \times m$ matrix of free parameters.

2. The probability function over words, expressed with C: a function g maps an input sequence of feature vectors for words in context, $(C(w_{t-n+1}), ..., C(w_{t-1}))$ and optionally the feature vector for the next word, $(C(w_{t-n+1}), ..., C(w_{t-1}))$ $C(w\_t)$ to a conditional probability distribution over words in V for the next word w_t. The output of g is a vector whose i-th element estimates the probability $\hat{P}(w_t = i \mid w_{t-n+1}^{t-1})$ as in Figure 6.1:

$$f(i, w_{t-1}, \cdots, w_{t-n+1}) = g(i, C(w_{t-1}), \cdots, C(w_{t-n+1})).$$

Alternatively, the energy minimization network used with some of the speeding up techniques described later (section 6.4) is based on

$$f(i, w_{t-1}, \cdots, w_{t-n+1}) = \frac{g(C(i), C(w_{t-1}), \cdots, C(w_{t-n+1}))}{\sum_j g(C(j), C(w_{t-1}), \cdots, C(w_{t-n+1}))}.$$

The function f is a composition of these two mappings (C and g), with C being *shared* across all the words in the context. With each of these two parts are associated some parameters. The parameters of the mapping C are simply the feature vectors themselves, represented by a $|V| \times m$ matrix C whose row i is the feature vector $C(i)$ for word i. The function g may be implemented by a feed-forward or recurrent neural network or another parameterized function, with parameters ω. The overall parameter set is $\theta = (C, \omega)$.

Training is achieved by looking for θ that maximizes the training corpus penalized log-likelihood:

$$L = \frac{1}{T} \sum_t \log f(w_t, w_{t-1}, \cdots, w_{t-n+1}; \theta) + R(\theta),$$

where $R(\theta)$ is a regularization term. For example, in our experiments, R is a weight decay penalty applied only to the weights of the neural network and to the C matrix, not to the biases.[2]

---

[2]The *biases* are the additive parameters of the neural network, such as b and d in equation 6.1 below.

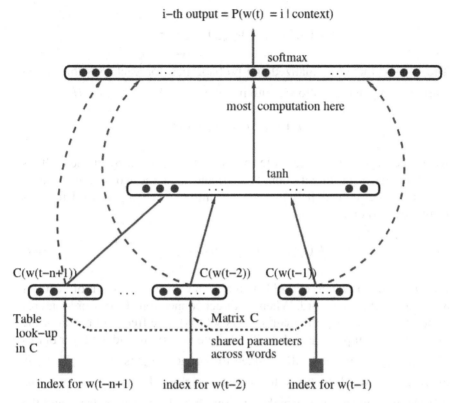

Fig 6.1 Neural architecture: $f(i, w_{t-1}, \cdots, w_{t-n+1}) = g(i, C(w_{t-1}), ..., C(w_{t-n+1}))$ where g is the neural network and $C(i)$ is the i-th word feature vector.

In the above model, the number of free parameters **only scales linearly** with $|V|$, the number of words in the vocabulary. It also **only scales linearly** with the order n: the scaling factor could be reduced to sub-linear if more sharing structure were introduced, e.g. using a time-delay neural network or a recurrent neural network (or a combination of both).

In most experiments below, the neural network has one hidden layer beyond the word features mapping, and optionally, direct connections from the word features to the output. Therefore there are really two hidden layers: the shared word features layer C, which has no non-linearity (it would not add anything useful), and the ordinary hyperbolic tangent hidden layer

More precisely, the neural network computes the following function, with a *softmax* output layer, which guarantees positive probabilities summing to 1:

$$\hat{P}(w_t \mid w_{t-1}, \cdots w_{t-n+1}) = \frac{e^{y_{w_t}}}{\sum_i e^{y_i}}.$$

The y_i are the unnormalized log-probabilities for each possible target word i, computed as follows, with parameters b, W, U, d and H:

$$y = b + Wx + U \tanh(d + Hx) \tag{6.1}$$

where the hyperbolic tangent (tanh) is applied element by element, W is optionally zero (no direct connections), and x is the word features layer activation vector, which is the concatenation of the input word features from the matrix C:

$$x = (C(w_{t-1}), C(w_{t-2}), \cdots, C(w_{t-n+1})). \tag{6.2}$$

Let h be the number of hidden units, and m the number of features associated with each word. When no direct connections from word features to outputs are desired, the matrix W is set to 0. The free parameters of the model are the output biases b (with $|V|$ elements), the hidden layer biases d (with h elements), the hidden-to-output weights U (a $|V| \times h$ matrix), the word features to output weights W (a $|V| \times (n-1)m$ matrix), the hidden layer weights H (an $h \times (n-1)m$ matrix), and the word features C (a $|V| \times m$ matrix):

$$\theta = (b, d, W, U, H, C).$$

The number of free parameters is $|V|(1 + nm + h) + h(1 + (n-1)m)$. The dominating factor is $|V|(nm + h)$. Note that in theory, if there is a weight decay on the weights W and H but not on C, then W and H could converge towards zero while C would blow up. In practice we did not observe such behavior when training with stochastic gradient ascent.

Stochastic gradient ascent on the neural network consists in performing the following iterative update after presenting the t-th word of the training corpus:

$$\theta \leftarrow \theta + \varepsilon \frac{\partial \log \hat{P}(w_t \mid w_{t-1}, \cdots w_{t-n+1})}{\partial \theta},$$

where ε is the "learning rate" (chosen as large as possible while allowing the average log-likelihood of the training set to continuously increase).

Note that a large fraction of the parameters needs not be updated or visited after each example: the word features $C(j)$ of all words j that do not occur in the input window.

Mixture of models. In our experiments (see Section 6.3) we have found improved performance by combining the probability predictions of the neural network with those of an interpolated n-gram model, either with a simple fixed weight of 0.5, a learned weight (maximum likelihood on the validation set) or a set of weights that are conditional on the frequency of the context (using the same procedure that combines trigram, bigram, and unigram in the interpolated trigram, which is a mixture).

6.3 First Experimental Results

Comparative experiments were performed on the Brown corpus which is a stream of 1,181,041 words, from a large variety of English texts and books. The first 800,000 words were used for training, the following 200,000 for validation (model selection: number of neurons, weight decay, early stopping) and the remaining 181,041 for testing. The number of different words is 47,578 (including punctuation, distinguishing between upper and lower case, and including the syntactical marks used to separate texts and paragraphs). Rare words with frequency ≤ 3 were merged into a single symbol, reducing the vocabulary size to $|V| = 16,383$.

An experiment was also run on text from the Associated Press (AP) News from 1995 and 1996. The training set is a stream of about 14 million (13,994,528) words, the validation set is a stream of about 1 million (963,138) words, and the test set is also a stream of about 1 million (963,071) words. The original data has 148,721 different words (including punctuation), which was reduced to $|V| = 17964$ by keeping only the most frequent words (and keeping punctuation), mapping upper case to lower case, mapping numeric forms to special symbols, mapping rare words to a special symbol and mapping proper nouns to another special symbol.

For training the neural networks, the initial learning rate was set to $\varepsilon_o = 10^{-3}$ (after a few trials with a tiny data set), and gradually decreased according to the following schedule: $\varepsilon_t = \frac{\varepsilon_o}{1+rt}$ where t represents the number of parameter updates done and r is a decrease factor that was heuristically chosen to be $r = 10^{-8}$.

6.3.1 Comparative Results

The first benchmark against which the neural network was compared is an interpolated or smoothed trigram model [Jelinek and Mercer, 1980]. Comparisons were also made with other state-of-the-art n-gram models: back-off n-gram models with the *Modified Kneser-Ney* algorithm [Kneser and Ney, 1995; Chen and Goodman, 1999], as well as class-based n-gram models [Brown *et al.*, 1992; Ney and Kneser, 1993; Niesler *et al.*, 1998]. The validation set was used to choose the order of the n-gram and the number of word classes for the class-based models. We used the implementation of these algorithms in the SRI Language Modeling toolkit, described by [Stolcke, 2002] and in www.speech.sri.com/projects/srilm/

In Tables 6.1 and 6.2 are measures of test set perplexity (geometric average of $1/\hat{P}(w_t \mid w_{t-n+1}^{t-1})$) for different models $\hat{P}$. Apparent convergence of the stochastic gradient ascent procedure was obtained after around 10 to 20 epochs for the Brown corpus. On the AP News corpus we were not able to see signs of over-fitting (on the validation set), possibly because we ran only 5 epochs. Early stopping on the validation set was used, but was necessary only in our Brown experiments. A weight decay penalty of 10^{-4} was used in the Brown experiments and a weight decay of 10^{-5} was used in the APNews experiments (selected by a few trials, based on validation set perplexity). Table 6.1 summarizes the results obtained on the Brown corpus. All the back-off models of the table are modified Kneser-Ney n-grams, which worked significantly better than standard back-off models. When c is specified for a back-off model in the table, a class-based n-gram is used (c is the number of word classes). Random initialization of the word features was done (similarly to initialization of neural network weights).

The **main result** is that significantly better results can be obtained when using the neural network, in comparison with the best of the n-grams, with a test perplexity difference of about 24% on Brown and about 8% on AP News, when taking the MLP versus the n-gram that worked best on the validation set. The table also suggests that the neural network was able to take advantage of more context (on Brown, going from 2 words of context to 4 words brought improvements to the neural network, not to the n-grams). It also shows that the hidden units are useful (MLP3 vs MLP1 and MLP4 vs MLP2), and that mixing the output probabilities of the neural network with the interpolated trigram always helps to reduce perplexity. The fact that simple averaging helps suggests that the neural network and the trigram make errors (i.e. low probability given to an observed word) in different places.

	n	c	h	m	direct	mix	Train	Valid	test
MLP1	5		50	60	yes	no	182	284	268
MLP2	5		50	60	yes	yes		275	257
MLP3	5		0	60	yes	no	201	327	310
MLP4	5		0	60	yes	yes		286	272
MLP5	5		50	30	yes	no	209	296	279
MLP6	5		50	30	yes	yes		273	259
MLP7	3		50	30	yes	no	210	309	293
MLP8	3		50	30	yes	yes		284	270
MLP9	5		100	30	no	no	175	280	276
MLP10	5		100	30	no	yes		265	252
Del. Int.	3						31	352	336
Kneser-Ney back-off	3							334	323
Kneser-Ney back-off	4							332	321
Kneser-Ney back-off	5							332	321
Class-based back-off	3	150						348	334
Class-based back-off	3	200						354	340
Class-based back-off	3	500						326	312
Class-based back-off	3	1000						335	319
Class-based back-off	3	2000						343	326
Class-based back-off	4	500						327	312
Class-based back-off	5	500						327	312

Table 6.1 Comparative results on the Brown corpus. The deleted interpolation trigram has a test perplexity that is 33% above that of the neural network with the lowest validation perplexity. The difference is 24% in the case of the best n-gram (a class-based model with 500 word classes). *n :* order of the model. *c:* number of word classes in class-based n-grams. *h:* number of hidden units. *m:* number of word features for MLPs. *direct*: whether there are direct connections from word features to outputs. *mix*: whether the output probabilities of the neural network are mixed with the output of the trigram (with a weight of 0.5 on each). The last three columns give perplexity on the training, validation and test sets.

	n	h	m	direct	Mix	Valid.	Test
MLP10	6	60	100	yes	yes	104	109
Del. Int.	3					126	132
Back-off KN	3					121	127
Back-off KN	4					113	119
Back-off KN	5					112	117

Table 6.2. Comparative results on AP News corpus. See table 1 for the column labels. KN stands for Kneser-Ney.

Table 6.2 gives similar results on the larger corpus (AP News), albeit with a smaller difference in perplexity (8%). Only 5 epochs were performed because of the computational load, but much more efficient implementations are described in the next few sections. The class-based model did not appear to help the n-gram models in this case, but the high-order modified Kneser-Ney back-off model gave the best results among the n-gram models. The MLP10 is mixed with the interpolated trigram, and it gave the best results (although better results might have been obtained by mixing with the back-off KN 5 model).

6.4 Architectural Extension: Energy Minimization Network

A variant of the above neural network can be interpreted as an energy minimization model following Hinton's work on products of experts [Hinton, 2000]. In the neural network described in the previous sections the distributed word features are used only for the "input" words and not for the "target" word (next word). Furthermore, a very large number of parameters (the majority) are expanded in the output layer: the semantic or syntactic similarities between target words are not exploited. In the variant described here, the target word is also represented by its feature vector. The network takes in input a sub-sequence of words (mapped to their feature vectors) and outputs an energy function E which is low when the words form a likely sub-sequence, high when it is unlikely. For example, the network outputs an "energy" function

$$E(w_{t-n+1}, \cdots, w_t) = v \cdot \tanh(d + Hx) + \sum_{i=0}^{n-1} b_{w_{t-i}} \qquad (6.3)$$

where b is the vector of biases (which correspond to unconditional probabilities), d is the vector of hidden units biases, v is the output weight vector, and H is the hidden layer weight matrix, and unlike in the previous model, input and target words contribute to x:

$$x = (C(w_t), C(w_{t-1}), C(w_{t-2}), \cdots, C(w_{t-n+1})).$$

The energy function $E(w_{t-n+1}, \cdots, w_t)$ can be interpreted as an unnormalized log-probability for the joint occurrence of $(w_{t-n+1}, \cdots, w_t)$. To obtain a conditional probability $\hat{P}(w_t \mid w_{t-n+1}^{t-1})$ it is enough (but costly) to normalize over the possible values of w_t with a *softmax*, as follows:

$$\hat{P}(w_t \mid w_{t-1}, \cdots, w_{t-n+1}) = \frac{e^{-E(w_{t-n+1}, \cdots, w_t)}}{\sum_i e^{-E(w_{t-n+1}, \cdots, w_{t-1}, i)}} .$$

Note that the total amount of computation is comparable to the architecture presented earlier, and the number of parameters can also be matched if the v parameter takes different values for each target word w_t. Note that only b_{w_t} remains after the above softmax normalization (any linear function of the w_{t-i} for $i > 0$ is canceled by the softmax normalization). As before, the parameters of the model can be tuned by stochastic gradient ascent on $\log \hat{P}(w_t \mid w_{t-1}, \cdots, w_{t-n+1})$, using similar computations.

In the products-of-experts framework, the hidden units can be seen as the experts: the joint probability of a sub-sequence $(w_{t-n+1}, \cdots, w_t)$ is proportional to the exponential of a sum of terms associated with each hidden unit j, $v_j \tanh(d_j + H_j x)$. Note that because we have chosen to decompose the probability of a whole sequence in terms of conditional probabilities for each element, the computation of the gradient is tractable. This is not the case for example with products-of-HMMs [Brown and Hinton, 2000], in which the product is over experts that view the whole sequence, and which can be trained with approximate gradient algorithms such as the contrastive divergence algorithm [Brown and Hinton, 2000]. Note also that this architecture and the products-of-experts formulation can be seen as extensions of the very successful **Maximum Entropy** models [Berger *et al.*, 1996], but where the basis functions (or "features", here the hidden units activations) are learned by penalized maximum likelihood at the same time as the parameters of the features linear combination, instead of being learned in an outer loop, with greedy feature subset selection methods.

In our experiments, we did not find significant generalization improvements with this architecture, but it lends itself more naturally to the speed-up techniques described below.

6.5 Speeding-up Training by Importance Sampling

In the above energy-based model, the probability distribution of a random variable X over some set X is expressed as

$$P(X = x) = \frac{e^{-\mathcal{E}(x)}}{Z} \qquad (6.4)$$

where $\mathcal{E}(\cdot)$ is a parameterized *energy* function which is low for plausible configurations of x, and high for improbable ones, and where $Z = \sum_{x \in X} e^{-\mathcal{E}(x)}$ is called the *partition function*. In the case that interests us, the partition function depends on the context $h_t = w_{t-n+1}^{t-1}$ because we estimate a conditional probability.

The main step in a gradient-based approach to train such models involves computing the gradient of the log-likelihood $\log P(X = x)$ with respect to the parameters θ of the energy function. The gradient can be decomposed in *two parts*: *positive reinforcement* for the observed value $X = x$ and *negative reinforcement* for every x', weighted by $P(X = x')$, as follows (by differentiating the negative logarithm of (4) with respect to θ):

$$\nabla_\theta \left(-\log P(x) \right) = \nabla_\theta \left(\mathcal{E}(x) \right) - \sum_{x' \in \chi} P(x') \nabla_\theta \left(\mathcal{E}(x') \right). \qquad (6.5)$$

Clearly, the difficulty here is to compute the negative reinforcement when $|\chi|$ is large (as is the case in a language modeling application). However, as is easily seen, the negative part of the gradient is nothing more than the average

$$E_P \left[\nabla_\theta \left(\mathcal{E}(X) \right) \right]. \qquad (6.6)$$

In [Hinton, 2002], it is proposed to estimate this average with a Gibbs sampling method, using a Markov Chain Monte-Carlo process. This technique relies on the particular form of the energy function in the case of products of experts, which lends itself naturally to Gibbs sampling (using the activities of the hidden units as one of the random variables, and the network input as the other one).

6.5.1 Approximation of the Log-Likelihood Gradient by Biased Importance Sampling

If one could sample from $P(\cdot)$, a simple way to estimate (Eq. 6.6) would consist in sampling M points $x_1, ..., x_M$ from the network's distribution $P(\cdot)$ and to approximate (Eq. 6.6) by the average

$$\frac{1}{M} \sum_{i=1}^{M} \nabla_\theta \left(\mathcal{E}(x_i) \right). \qquad (6.7)$$

This method, known as *classical Monte-Carlo* allows estimating the gradient of the log-likelihood (Eq. 6.5). The maximum speed-up that could be achieved with such a procedure would be $|\chi|/M$. In the case of the language modeling application we are considering, that means a potential for a huge speed-up, since $|\chi|$ is typically in the tens of thousands and M could be quite small; in fact, Hinton found $M=1$ to be a good choice with the contrastive divergence method [Hinton, 2002].

However, this method requires to sample from distribution $P(\cdot)$, but it is not clear how to do this efficiently without computing $\mathcal{E}(x)$ for each $x \in \chi$.

Fortunately, in many applications, such as language modeling, we can use an alternative, *proposal* distribution Q from which it is cheap to sample. In the case of language modeling, for instance, we can use n-gram models. There exist several Monte-Carlo algorithms that can take advantage of such a distribution to give an estimate of (Eq. 6.6).

Classical Importance Sampling

One well-known statistical method that can make use of a proposal distribution Q in order to approximate the average $E_P[\nabla_\theta(\mathcal{E}(X))]$ is based on a simple observation. In the discrete case

$$E_P[\nabla_\theta(\mathcal{E}(x))] = \sum_{x \in \chi} P(x) \nabla_\theta(\mathcal{E}(x))$$

$$= \sum_{x \in \chi} Q(x) \frac{P(x)}{Q(x)} \nabla_\theta(\mathcal{E}(x)) = \mathcal{E}_Q \left[\frac{P(X)}{Q(X)} \nabla_\theta(\mathcal{E}(X)) \right].$$

Thus, if we take M independent samples $x_1, ..., x_M$ from Q and apply classical Monte-Carlo to estimate $E_Q \left[\frac{P(X)}{Q(X)} \nabla_\theta(\mathcal{E}(X)) \right]$, we obtain the following estimator known as *importance sampling* [Robert and Casella, 2000]:

$$\frac{1}{M} \sum_{i=1}^{M} \frac{P(x_i)}{Q(x_i)} \nabla_\theta(\mathcal{E}(x_i)). \tag{6.8}$$

Clearly, that does not solve the problem: although we do not need to sample from P anymore, the $P(x_i)$'s still need to be computed, which cannot be done without explicitly computing the partition function. Back to square one.

Biased Importance Sampling

Fortunately, there is a way to estimate (Eq. 6.6) without sampling from P nor having to compute the partition function. The proposed estimator is a biased version of classical importance sampling [Kong *et al.*, 1994]. It can be used when $P(x)$ can be computed explicitly up to a multiplicative constant: in the case of energy-based models, this is clearly the case since $P(x) = Z^{-1}e^{-\mathcal{E}(x)}$. The idea is to use $\frac{1}{W}w(x_i)$ to weight the $\nabla_\theta(\mathcal{E}(x_i))$, with $w(x) = \dfrac{e^{-\mathcal{E}(x)}}{Q(x)}$ and $W = \sum_{j=1}^{M} w(x_j)$, thus yielding the estimator [Liu, 2001]

$$\frac{1}{W}\sum_{i=1}^{M} w(x_i)\nabla_\theta\big(\mathcal{E}(x_i)\big). \tag{6.9}$$

Though this estimator is biased, its bias decreases as M increases. It can be shown to converge to the true average (Eq. 6.6) as $M \to \infty$ [3].

The advantage of using this estimator over classical importance sampling is that we no more need to compute the partition function: we just need to compute the energy function for the sampled points.

Adapting the Sample Size

Preliminary experiments with biased importance sampling using the unigram distribution showed that whereas a small sample size was appropriate in the initial training epochs, a larger sample size was necessary later to avoid divergence (increasing training error). This may be explained by a too large bias – because the network's distribution diverges from that of the unigram, as training progresses – and/or by a too large variance in the gradient estimator.

In [Bengio and Senécal, 2003], we presented an improved version of biased importance sampling for the neural language model that makes use of a diagnostic, called *effective sample size* [Kong, 1992; Kong *et al.*, 1994]. For a sample $x_1,...,x_M$ taken from proposal distribution Q, the effective sample size is defined by

$$ESS = \frac{\left(\sum_{j=1}^{M} w(x_j)\right)^2}{\sum_{j=1}^{M} w(x_j)^2} \tag{6.10}$$

---

[3]However, this does not guarantee that the variance of the estimator remains bounded. We have not dealt with the problem yet, but see [Luis and Leslie, 2000].

Basically, this measure approximates the number of samples from the target distribution P that would have yielded, with classical Monte-Carlo, the same variance as the one yielded by the biased importance sampling estimator with sample $x_1, ..., x_m$.

We can use this measure to diagnose whether we have sampled enough points. In order to do that, we fix a baseline sample size l. This baseline is the number of samples we would sample in a classical Monte-Carlo scheme, were we able to do it. We then sample points from Q by "blocks" of size $m_b \geq 1$ until the effective sample size becomes larger than the target l. If the number of samples becomes too large, we switch back to a full back-propagation (i.e. we compute the true negative gradient).

Adapting the Proposal Distribution

The method was used with a simple unigram proposal distribution to yield significant speed-up on the Brown corpus [Bengio and Senécal, 2003]. However, the required number of samples was found to increase quite drastically as training progresses. This is because the unigram distribution stays fixed while the network's distribution changes over time and becomes more and more complex, thus diverging from the unigram. Switching to a bigram or trigram during training actually worsens even more the training, requiring even larger samples.

Clearly, using a proposal distribution that stays "close" to the target distribution would yield even greater speed-ups, as we would need less samples to approximate the gradient. We propose to use a n-gram model that is *adapted* during training to fit to the target (neural network) distribution P[4]. In order to do that, we propose to *redistribute the probability mass* of the sampled points in the n-gram to track P. This is achieved with n-gram tables Q_k which are estimated with the goal of matching the order k conditional probabilities of samples from our model P when the history is sampled from the empirical distribution. The tables corresponding to different values of k are interpolated in the usual way to form a predictive model, from which it is easy to sample.

Let us thus define the *adaptive n-gram* as follows:

$$Q(w_t \mid h_t) = \sum_{k=1}^{n} \alpha_k(h_t) Q_k(w_t \mid w_{t-k+1}^{t-1}) \qquad (6.11)$$

where $h_t = w_{t-n+1}^{t-1}$ is the context, the Q_k are the sub-models that we wish

---

[4]A similar approach was proposed in [Cheng and Druzdzel, 2000] for Bayesian networks.

to estimate, and $\alpha_k(h_t)$ is a mixture function such that $\sum_{k=1}^{n} \alpha_k(h_t) = 1$. Usually, for obvious reasons of memory constraints, the probabilities given by an n-gram are non-null only for those sequences that are observed. Mixing with lower-order models allows to give some probability mass to unseen word sequences.

Let W be the set of M words sampled from Q. Let $\bar{q}_k = \sum_{w \in W} Q_k(w \mid w_{t-k+1}^{t-1})$ be the total probability mass of the sampled points in k-gram Q_k and $\bar{p} = \sum_{w \in W} e^{-\mathcal{E}(w,h_t)}$ the unnormalized probability mass of these points in P. Let $\tilde{P}(w \mid h_t) = \frac{e^{-E(w,h_t)}}{\bar{p}}$ for each $w \in W$. For each k and for each $w \in W$, the values in Q_k are updated as follows:

$$Q_k(w \mid w_{t-k+1}^{t-1}) \leftarrow (1 - \lambda)Q_k(w \mid w_{t-k+1}^{t-1}) + \lambda \bar{q}_k \tilde{P}(w \mid h_t) \qquad (6.12)$$

where λ is a kind of "learning rate". The parameters of functions $\alpha_k(\cdot)$ are updated so as to minimize the Kullback-Leibler divergence $\sum_{w \in W} \tilde{P}(w \mid h_t) \log \frac{\tilde{P}(w \mid h_t)}{Q(w \mid h_t)}$ by gradient descent.

We describe here the method we used to train the α_k's in the case of a bigram interpolated with a unigram, i.e. $n = 2$ above. In our experiments, the α_k's were a function of the frequency of the last word w_{t-1}.

The words were first clustered in C frequency bins $B_c, c = 1, ..., C$. Those bins were built so as to *group words with similar frequencies in the same bin while keeping the bins balanced*[5].

Then, an "energy" value $a(c)$ was assigned for $c = 1, ..., C$. We set $\alpha_1(h_t) = \sigma(a(h_t))$ and $\alpha_2(h_t) = 1 - \alpha_1(h_t)$ where $\sigma(z) = 1/(1 + e^{-z})$ is the sigmoid function and $a(h_t) = a(w_{t-1}) = a(c)$, c being the class (bin) of w_{t-1}. The energy $a(h_t)$ is thus updated with the following rule:

---

[5]By *balanced*, we mean that the sum of word frequencies does not vary a lot between two bins. That is, let $|w|$ be the frequency of word $|w|$ in the training set, then we wish that $\forall i, j, \sum_{w \in B_i} |w| \approx \sum_{w \in B_j} |w|$

$$a(h_t) \leftarrow a(h_t) - \eta \alpha_1(h_t) \alpha_2(h_t) \sum_{w \in W} \tilde{P}(w \mid h_t) \frac{Q(w \mid h_t)}{Q_2(w \mid w_{t-1}) - Q_1(w)}$$

$$(6.13)$$

where η is a learning rate.

6.5.2 Experimental Results

We ran some experiments on the Brown corpus, with different configurations. For these experiments the vocabulary was truncated by mapping all "rare" words (words that appear 3 times or less in the corpus) into a single special word. The resulting vocabulary contains 14,847 words. The preprocessing being a bit different, the numerical results are not directly comparable with those of Section 6.3.1.

On this dataset, a simple interpolated trigram, serving as our baseline, achieves a perplexity of 253.8 on the test set [6]. In all settings, we used 30 word features for both context and target words, and 80 hidden neurons. The number of context words was 3. Initial learning rate for the neural network was set to 3.0^{-3}, decrease constant to 10^{-8} and we used a weight decay of 10^{-4}. The output biases b_{w_t} in (3) were manually initialized so that the neural network's initial distribution is equal to the unigram. This setting is the same as that of the neural network that achieved the best results on Brown, as described in Section 6.3.1. In this setting, a classical neural network – one that doesn't make a sampling approximation of the gradient – converges to a perplexity of 204 in test, after 18 training epochs. In the adaptive bigram algorithm, the parameters λ in (Eq. 6.12) and η in (Eq. 6.13) where both set to 10^{-3}. The $a(h_t)$ were initially set to $\sigma^{-1}(0.9)$ so that $\alpha_1(h_t) = 0.9$ and $\alpha_2(h_t) = 0.1$ for all h_t; this way, at the start of training, the target (neural network) and the proposal distribution are close to each other (both are close to the unigram).

---

[6]Better results can be achieved with a Knesser-Ney back-off trigram, but it has been shown before that a neural network converges to a lower perplexity on Brown. Furthermore, the neural network can be interpolated with the trigram for even larger perplexity reductions.

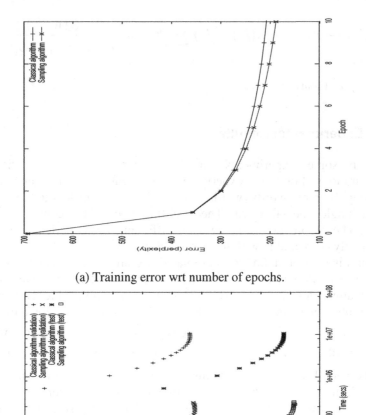

(a) Training error wrt number of epochs.

. (b) validation and test errors wrt CPU time

Fig 6.2 Comparison of errors between a model trained with the classical
algorithm and a model trained by adaptive importance sampling

Figure 6.2(a) plots the training error at every epoch for the network
trained without sampling (Section 6.4) and a network trained by
importance sampling, using an adaptive bigram with a target effective
sample size of 50. The number of frequency bins used for the mixing
variables was 10. It shows that the convergence of both networks is
similar. The same holds for validation and test errors, as is shown in
Figure 6.2(b). In this figure, the errors are plotted wrt computation time on

a Pentium 4 2 GHz. As can be seen, **the network trained with the sampling approximation converges before the network trained classically even finishes to complete one full epoch**.

Quite interestingly, the network trained by sampling converges to an even lower perplexity than the ordinary one (trained with the exact gradient). After 9 epochs (26 hours), its perplexity over the test set is equivalent to that of the one trained with exact gradient at its overfitting point (18 epochs, 113 days). The sampling approximation thus allowed a **100-fold speed-up.**

Surprisingly enough, if we let the sampling-trained model converge, it starts to overfit at epoch 18, as for classical training, but with a lower test perplexity of 196.6, a 3.8% improvement. Total improvement in test perplexity with respect to the trigram baseline is 29%.

An important detail is worth mentioning here. Since $|V|$ is large, we first thought that there was too small a chance to sample the same word w twice from the proposal Q at each step to really worry about it. However, we found out the chance of picking twice the same w to be quite high in practice (with an adaptive bigram proposal distribution). We think this is due to the particular look of our proposal distribution (Eq. 6.11). The bigram part $Q_2(w \mid w_{t-1})$ of that distribution being non-null for only those words w for which $|w_{t-1}w| > 0$, there are contexts w_{t-1} in which the number of candidate words w for which $Q_2(w \mid w_{t-1}) > 0$ is small, thus there are actually good chances to pick twice the same word. Knowing this, one can save much computation time by avoiding computing the energy function $\mathcal{E}(w, h_t)$ many times for the same word w. Instead, the values of the energy functions for sampled words is kept in memory. When a word is first picked, its energy $\mathcal{E}(w, h_t)$ is computed in order to calculate the sampling weights. The value of the sampling weight is kept in memory so that, whenever the same word is picked during the same iteration, all that needs to be done is to use the copied weight, thus saving one full propagation of the energy function. **This trick increases the speed-up from a factor of 100 to a factor of 150.**

6.6 Speeding-up Probability Computation by Hierarchical Decomposition

In this section we consider a very fast variant of the neural probabilistic language model, in which not just the training algorithm but also the model itself is different. It is based on an idea that could in principle deliver close to exponential speed-up with respect to the number of words in the vocabulary. The computations required during training and during probability prediction with the regular model are a small constant plus a factor linearly proportional to the number of words $|V|$ in the vocabulary V. The approach proposed here can yield a speed-up of order $O\left(\frac{|V|}{\log|V|}\right)$ for the second term. It follows up on a proposal made in [Goodman, 2001b] to rewrite a probability function based on a partition of the set of words. The basic idea is to form a hierarchical description of a word as a sequence of $O(\log|V|)$ decisions, and to learn to take these probabilistic decisions instead of directly predicting each word's probability. Another important idea presented here is to reuse the same model (i.e. the same parameters) for all those decisions (otherwise a very large number of models would be required), using a special symbolic input that characterizes the nodes in the tree of the hierarchical decomposition. Finally, we use prior knowledge in the WordNet lexical reference system to help define the hierarchy of word classes.

6.6.1. Hierarchical Decomposition Can Provide Exponential Speed-up

In [Goodman, 2001b] it is shown how to speed-up a maximum entropy class-based statistical language model by using the following idea. Instead of computing directly $P(Y|X)$ (which involves normalization across all the values that Y can take), one defines a clustering partition for the Y (into the word classes C, such that there is a deterministic function from Y to C), so as to write

$$P(Y = y \mid X = x) = P(Y = y \mid C = c(y), X = x)P(C = c(y) \mid X = x).$$

This is always true for any function $c(y)$ because

$$P(Y \mid X) = \sum_c P(Y, C = c \mid X) = \sum_c P(Y \mid C = c, X)P(C = c \mid X)$$
$$= P(Y \mid C = c(Y), X)P(C = c(Y) \mid X)$$

since only one value of C is compatible with the value of Y. However, generalization could be better for choices of word classes that "make sense", i.e. make it easier to learn the $P(C = c(y) \mid X = x)$. If Y can take 10000 values and we have 100 classes with 100 words y in each class, then instead of doing normalization over 10000 choices we only need to do two normalizations, each over 100 choices. If computation of conditional probabilities is proportional to the number of choices then the above would reduce computation by a factor 50. This is approximately what is gained according to the measurements reported in [Goodman, 2001b]. The same paper suggests that one could introduce more levels to the decomposition and here we push this idea to the limit. Indeed, whereas a one-level decomposition should provide a speed-up on the order of $\frac{|V|}{\sqrt{|V|}} = \sqrt{|V|}$, a hierarchical decomposition represented by a balanced binary tree should provide an exponential speed-up, on the order of $\frac{|V|}{\log_2 |V|}$ (at least for the part of the computation that is linear in the number of choices).

Each word v must be represented by a bit vector $(k_1(v),...k_m(v))$ (where m depends on v). This can be achieved by building a binary hierarchical clustering of words, and a method for doing so is presented in the next section. For example, $k_1(v) = 1$ indicates that v belongs to the top-level group 1 and $k_2(v) = 0$ indicates that it belongs to the sub-group 0 of that top-level group.

The next-word conditional probability can thus be represented and computed as follows:

$$P(v \mid w_{t-1},...) = P(k_1(v) \mid w_{t-1},...)P(k_2(v) \mid k_1(v), w_{t-1},...)$$
$$... P(k_m(v) \mid k_1(v),...k_{m-1}(v), w_{t-1},...). \qquad (6.14)$$

This can be interpreted as a series of binary decisions associated with nodes of a binary tree. Each node is indexed by a bit vector corresponding to the path from the root to the node (append 1 or 0 according to whether the left or right branch of a decision node is followed). Each leaf corresponds to a word. If the tree is balanced then the maximum length of the bit vector is $\lceil \log_2 |V| \rceil$. Note that we could further reduce computation by looking for an encoding that takes the frequency of words into account, to reduce the average bit length to the unconditional entropy of words. For example with the corpus used in our experiments, $|V| = 10000$ so $\log_2 |V| \approx 13.3$ while the unigram entropy is about 9.16, i.e. there is a possible speed-up of 31%. The gain would be greater for larger vocabularies, but not a very significant improvement over the major one obtained by using a simple balanced hierarchy.

The "target class" (0 or 1) for each node is obtained directly from the target word in each context, using the bit encoding of that word. Note also that there is a target (and gradient propagation) only for the nodes on the path from the root to the leaf associated with the target word. This is the major source of savings during training.

During recognition and testing, there are two main cases to consider: (1) one needs the probability of only one word, e.g. the observed word, (or very few), or (2) one needs the probabilities of all the words. In the first case (which occurs during testing on a corpus) we still obtain the exponential speed-up. In the second case, we are back to $O(|V|)$ computations (with a constant factor overhead). For the purpose of estimating generalization performance (out-of-sample log-likelihood) only the probability of the observed next word is needed. And in practical applications such as speech recognition, we are only interested in discriminating between a few alternatives, e.g. those that are consistent with the acoustics, and represented in a trellis of possible word sequences.

This speed-up should be contrasted with the one provided by the importance sampling method described above (Section 6.5), which leads to significant speed-up during training, but because the architecture is unchanged, probability computation during recognition and test still requires $O(|V|)$ computations for each prediction. Instead, the architecture proposed here gives significant speed-up both during training and test / recognition.

6.6.2 Sharing Parameters Across the Hierarchy

If a separate predictor is used for each of the nodes in the hierarchy, about $2|V|$ predictors are needed. This represents a huge capacity since each predictor maps from the context words to a single probability. This might create problems in terms of computer memory (not all the models would fit at the same time in memory) as well as overfitting. Therefore we have chosen to build a model in which parameters are shared across the hierarchy. There are clearly many ways to achieve such sharing, and alternatives to the architecture presented here should motivate further study.

Based on our discussion in the introduction, it makes sense to force the word embedding to be shared across all nodes. This is important also because the matrix of word features C is the largest component of the parameter set. Since each node in the hierarchy has a semantic meaning (being associated with a group of hopefully similar-meaning words) it makes sense to also associate each node with a feature vector. Without loss of generality, we can consider the model to predict $P(k \mid node, w_{t-1}, ..., w_{t-n+1})$, where *node* corresponds to a sequence of bits specifying a node in the hierarchy and k is the next bit (0 or 1), corresponding to one of the two children of *node*. This can be represented by a model similar to the one described in Section 6.2, but with two kinds of symbols in input: the context words and the current node. We allow the embedding parameters for word cluster nodes to be different from those for words. Otherwise the architecture is the same, with the difference that there are only two choices to predict, instead of $|V|$ choices.

More precisely, the specific predictor used in our experiments is the following:

$$P(k = 1 \mid node, w_{t-1}, ..., w_{t-n+1}) = \sigma(\gamma_{node} + \beta \cdot \tanh(d + Hx + UN_{node}))$$

where x is the concatenation of context word features as in Eq. 6.2, σ is the sigmoid function, γ_{node} is a bias parameter playing the same role as b_{w_t} in Eq. 6.3, β is a weight vector playing the same role as v in Eq. 6.3, d and H play the same role as in Eq. 6.3, and N gives feature vector embeddings for nodes, with U playing the same role as H for the node feature vector embedding.

6.6.3 Using WordNet to Build the Hierarchical Decomposition

A very important component of the whole model is the choice of the words binary encoding, i.e. of the hierarchical word clustering. In this paper we combine empirical statistics with prior knowledge from the WordNet resource [Fellbaum, 1998]. Again there are many other choices that could have been made, one extreme being a purely data-driven clustering of words.

The IS-A taxonomy in WordNet organizes semantic concepts associated with senses in a graph that is almost a tree. For our purposes we need a tree, so we have manually selected a parent for each of the few nodes that have more than one. The leaves of the WordNet taxonomy are senses and each word can be associated with more than one sense. Words sharing the same sense are considered to be synonymous (at least in one of their uses). For our purpose we have to choose one of the senses for each word (to transform the hierarchy over senses into a hierarchy over words) and we selected the most frequent sense. In addition, this WordNet tree is not binary: each node may have many more than two children (this is particularly a problem for verbs and adjectives, for which WordNet is shallow and incomplete). To transform this hierarchy into a binary tree we perform a binary hierarchical clustering of the children associated with each node, as illustrated in Figure 6.3. The K-means algorithm is used at each step to split each cluster. To compare nodes, we associate each node with the subset of words that it covers. Each word is associated with a TF/IDF [Salton and Buckley, 1988] vector of document/word occurrence counts, where each "document" is a paragraph in the training corpus. Each node is associated with the dimension-wise median of the TF/IDF scores. Each TF/IDF score is the occurrence frequency of the word in the document times the logarithm of the ratio of the total number of documents by the number of documents containing the word.

6.6.4 Comparative Results

Experiments were performed to evaluate the speed-up and any change in generalization error. The experiments also compared an alternative speed-up technique described in Section 6.5, based on importance sampling. The experiments were performed on the Brown corpus, with a reduced vocabulary size of 10,000 words (the most frequent ones). The corpus was split into 3 sets: 900,000 for training, 100,000 for validation (model selection), and 105,515 for testing. Again, the absolute perplexity numbers cannot be directly compared to those given earlier. The validation set was used to select a small number of choices for the size of the embeddings and the number of hidden units.

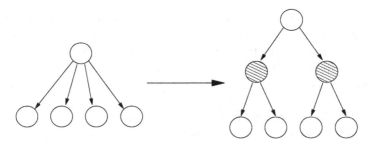

Fig 6.3 WordNet's IS-A hierarchy is not a binary tree: most nodes have many children. Binary hierarchical clustering of these children is performed.

The results in terms of raw computations (time to process one example), either during training or during test are shown respectively in Tables 6.3 and 6.4. The computations were performed on Athlon processors with a 1.2 GHz clock. The speed-up during training is by a factor greater than 250 and during test by a factor close to 200. These are impressive but less than the $|V| / \log_2 |V| \approx 750$ that could be expected if there was no overhead and no constant term in the computational cost.

Architecture	Time per epoch (set)	Time per example (ms)	Speed-up
Original neural net	416 300	462.6	1
Importance sampling	6 092	6.73	68.7
Hierarchical model	1 609	1.79	258

Table 6.3 Training time per epoch (going once through all the training examples) and per example. The original neural net is as described in section 6.2 The importance sampling algorithm trains the same model faster. The hierarchical model is the one proposed here.

Architecture	Time per example (ms)	Speed-up
Original neural net	270.7	1
Importance sampling	221.3	1.22
Hierarchical model	1.4	193

Table 6.4 Test time per example for the different algorithms. See Table 6.3's caption.

It is also important to verify that learning still works and that the model generalizes well. Training is performed over about 20 to 30 epochs according to validation set perplexity (early stopping). Table 6.5 shows the comparative generalization performance of the different architectures, along with that of an interpolated 3-gram (same procedure as in [Bengio *et al.*, 2003], which follows [Jelinek and Mercer, 1980]). Note that better performance should be obtainable with some of the tricks in [Goodman, 2001a]. Combining the neural network with a trigram should also decrease its perplexity, as already shown earlier (see Section 6.3.1).

	Validation perplexity	Test Perplexity
Interpolated trigram	299.4	268.7
Original neural net	213.2	195.3
Importance sampling	209.4	192.6
Hierarchical model	241.6	220.7

Table 6.5 Test perplexity for the different architectures and for an interpolated trigram.

As shown in Table 6.5, the hierarchical model does not generalize as well as the original neural network, but the difference is not very large and still represents an improvement over the basic interpolated trigram model. Given the very large speed-up, it is certainly worth investigating variations of the hierarchical model proposed here (in particular how to define the hierarchy) for which generalization could be better. Note also that the speed-up would be greater for larger vocabularies (e.g. 50,000 is not uncommon in speech recognition systems).

6.7 Short Lists and Speech Recognition Applications

The standard measure of the prediction capability of language models is perplexity, as used in our experiments described above. Often language models are however part of a larger system and improvements in perplexity do not necessarily lead to better performance of the overall system. An important application domain of language models is continuous speech recognition and many new language modeling techniques have been in fact first introduced in the context of this domain.

In continuous speech recognition we are faced with the problem to find the word sequence w^* corresponding to a given acoustic signal x. Using Bayes' rule,

$$w^* = \operatorname*{argmax}_{w} \hat{P}(w \mid x)$$

$$= \operatorname*{argmax}_{w} \frac{f(x \mid w)\hat{P}(w)}{P(x)}$$

$$= \operatorname*{argmax}_{w} f(x \mid w)\hat{P}(w)$$

$$(6.15)$$

where $f(x \mid w)$ is the so called acoustic model and $\hat{P}(w)$ the language model. The problem of actually finding w^* that maximizes the equation is known as the decoding problem. To the best of our knowledge, all state-of-the-art large vocabulary continuous speech recognizers (LVCSR) used hidden Markov models (with various methods for parameter sharing and adaptation) for acoustic modeling and n-gram back-off language models to estimate $\hat{P}(w)$.

In the following we study the application of the neural network language model to conversational telephone speech recognition in a

LVCSR. Recognition of conversational speech is a very challenging tasks with respect to acoustic modeling (bad signal quality, highly varying pronunciations, ...) as well as language modeling (unconstrained speaking style, frequent grammatical errors, hesitations, start-overs, ..). This is illustrated by the following examples that have been extracted from the training material:

- *But yeah so that is that is what I do.*
- *I I have to say I get time and I get a few other.*
- *Possibly I don't know I mean I had one years ago so.*
- *Yeah yeah well it's it's.*

In addition, language modeling for conversational speech suffers from a lack of adequate training data since the main source is audio transcriptions, in contrast to the broadcast news task for which other news sources are readily available. Unfortunately, collecting large amounts of conversational data and producing detailed transcriptions is very costly. One possibility is to increase the amount of training data by selecting conversational-like sentences in broadcast news material and on the Internet, or by transforming other sources to be more conversational-like. Here we use the neural network language model in order to see if it can take better advantage from the limited amount of training data. Initial work has in fact shown that the neural network language model (LM) can be used to reduce the word error rate in a speech recognizer [Schwenk and Gauvain, 2002]. These results were later confirmed with an improved speech recognizer [Gauvain *et al.*, 2002; Schwenk and Gauvain, 2003]. A neural network LM has also been used with a Syntactical Language Model showing perplexity and word error improvements on the Wall Street Journal corpus [Emami *et al.*, 2003].

In our previous work the neural network LM was trained on about 6M words and the word error rates were in the range of 25%. Since then, large amounts of conversational data have been collected in the DARPA EARS program and quick transcriptions were made available (total of about 2200h, 27.2M words), helping to build better speech recognizers. This amount of LM training data enables better training of the back-off 4-gram LM and one may wonder if the neural network LM, a method developed for probability estimation of sparse data, still achieves an improvement. In the following sections we show that the neural network LM continues to achieve consistent word error reductions with respect to a carefully tuned back-off 4-gram LM. Detailed results are provided as a function of the size of the LM training data. The described systems have successfully participated in speech recognition evaluations organized by the

National Institute of Standards and Technology NIST and DARPA [Lee *et al.*, 2003; Fiscus *et al.*, 2004]. All results reported in the following sections use the official test sets of these evaluations.

Incorporating this approach into a heavily tuned LVCSR needs however several modifications of the basic architecture described above. Current decoding algorithms suppose in particular that a request of a LM probability takes basically no time[7] and many thousand LM probabilities may be requested to decode a sentence of several seconds. Fast training is also important for system development. The necessary modifications of the neural network language model are described in the following sections. We first present another approach for fast training of the neural networks and efficient incorporation into a speech recognition system [Schwenk, 2004]. We then summarize the principles of the reference speech recognizer.

6.7.1 Fast Recognition

Language models play an important role during decoding of continuous speech since the information provided about the most probable set of words given the current context is used to limit the search space. Using the neural LM directly during decoding imposes an important burden on search space organization since a context of three words must be kept. This led to long decoding times in our first experiments when the neural LM was used directly during decoding [Schwenk and Gauvain, 2002]. In order to make the model tractable for LVCSR the following techniques have been applied (more details below):

1. **Lattice rescoring**: decoding is done with a standard back-off LM and a lattice is generated. The neural network LM is then used to rescore the lattice, i.e. the LM probabilities are changed.
2. **Shortlists**: the neural network is only used to predict the LM probabilities of a subset of the whole vocabulary.
3. **Regrouping**: all LM probability requests in one lattice are collected and sorted. By these means all LM probability requests with the same context h_t lead to only one forward pass through the neural network.
4. **Block mode**: several examples are propagated at once through the neural network, allowing the use of faster matrix/matrix operations.

---

[7] For a back-off LM this is a table look-up using hashing techniques.

5. **CPU optimization**: machine specific BLAS libraries are used for fast matrix and vector operations.

Normally, the output of a speech recognition system is the most likely word sequence given the acoustic signal (see Eq. 6.15), but it is often advantageous to preserve more information for subsequent processing steps. This is usually done by generating a lattice which is a graph of possible solutions where each edge corresponds to a hypothesized word with its acoustic and language model scores. These graphs are called lattices (see Figure 6.4 for an example). In the context of this work, the lattice is used to replace the LM probabilities by those calculated by the neural network LM (*lattice rescoring*).

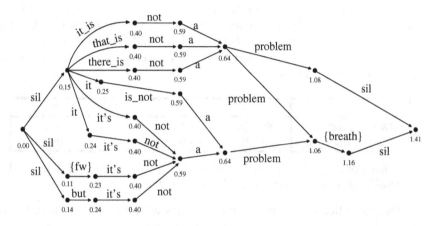

Fig 6.4 Example of a lattice produced by the speech recognizer using a 3-gram language model. A lattice is a graph of possible solutions where each edge corresponds to a hypothesized word with its acoustic and language model scores (for clarity these scores are not shown in the figure). {*fw*} stands for a filler word and {*breath*} is breath noise.

It has been demonstrated in Section 6.2 that most calculations are due to the large size of the output layer. Remember that all outputs need to be calculated in order to perform the softmax normalization even though only one LM probability is needed. Experiments using lattice rescoring with normalized LM scores led to much higher word error rates. One may argue that it is not very reasonable to spend a lot of time to get the LM probabilities of words that do not appear very often. Therefore, we chose to limit the output of the neural network to the s most frequent words, $s = |V|$, referred to as a *shortlist* in the following discussion. All words in V are still considered for the input of the neural network. The LM probabilities of words in the shortlist ($\hat{P}_N$) are calculated by the neural network and the LM probabilities of the remaining words ($\hat{P}_B$) are obtained from a standard 4-gram back-off LM:

$$\hat{P}(w_t \mid h_t) = \begin{cases} \hat{P}_N(w_t \mid h_t) \cdot P_S(h_t) & \text{if } w_t \in \text{shortlist} \\ \hat{P}_B(w_t \mid h_t) & \text{else} \end{cases}$$

$$P_S(h_t) = \sum_{w \in shortlist(h_t)} \hat{P}_B(w \mid h_t)$$

It can be considered that the neural network redistributes the probability mass of all the words in the shortlist[8]. This probability mass is precalculated and stored in the data structures of the standard 4-gram LM. A back-off technique is used if the probability mass for a requested input context is not directly available. Table 6.6 gives the coverage, i.e. the percentage of LM probabilities that are effectively calculated by the neural network when evaluating the perplexity on a development set of 56k words ot when rescoring lattices.

Shortlist size	1024	2000	4096	8192
Development set	89.3%	93.6%	96.8%	98.5%
Lattice rescoring	88.5%	89.9%	90.4%	91.0%

Table 6.6 Coverage for different shortlist sizes, i.e. percentage of 4-grams that are actually calculated by the neural LM. The vocabulary size is about 51k.

During lattice rescoring LM probabilities with the same context h_t are often requested several times on potentially different nodes in the lattice (for instance the trigram "*not a problem*" in the lattice shown in Figure 6.4). Collecting and regrouping all these calls prevents multiple forward passes since all LM predictions for the same context are immediately available (see Figure 6.1). Further improvements can be obtained by propagating several examples at once through the network, which is also known as bunch mode [Bilmes *et al.*, 1997]. In comparison to equation 1, this results in using matrix/matrix instead of matrix/vector operations:

---

[8]Note that the sum of the probabilities of the words in the shortlist for a given context is normalized $\sum_{w \in shortlist} \hat{P}_N(w \mid h_t) = 1$.

$$Z = \tanh\left(HX + D\right)$$
$$Y = UZ + B$$

where D and B are obtained by duplicating the bias d and b respectively for each line of the matrix. The tanh-function is performed element-wise. These matrix/matrix operations can be aggressively optimized on current CPU architectures, e.g. using SSE2 instructions for Intel processors [Intel's MKL, 2004; ATLAS, 2004]. Although the number of floating point operations to be performed is strictly identical to single example mode, an up to five times faster execution can be observed depending on the sizes of the matrices.

The test set of the 2001 NIST speech recognition evaluation consists in 6h of speech comprised of 5895 conversations sides. The lattices generated by the speech recognizer for this test set contain on average 511 nodes and 1481 arcs per conversation side. In total 3.8 million 4-gram LM probabilities were requested out of which 3.4 million (89.9%) have been processed by the neural network, i.e. the word to be predicted is among the 2000 most frequent words (the shortlist). After collecting and regrouping all LM calls in each lattice, only 1 million forward passes though the neural network have been performed, giving a cache hit rate of about 70%. Using a bunch size of 128 examples, the total processing time took less than 9 minutes on a Intel Xeon 2.8GHz processor, i.e. in 0.03 times real time (xRT).[9] This corresponds to about 1.7 billion floating point operations per second (1.7 GFlops). Lattice rescoring without bunch mode and regrouping of all calls in one lattice is about ten times slower.

6.7.2 Fast Training

Language models for LVCSR are usually trained on text corpora of several million words. With a vocabulary size of 51k words, standard back-propagation training would take a very long time. In addition to the speed-up techniques of importance sampling and hierarchical decomposition described in the previous sections, a third method has been developed. Optimized floating point operations are much more efficient if they are applied to data that is stored in contiguous locations in memory, making a better use of cache and data prefetch capabilities of processors. This is

---

[9]In speech recognition, processing time is measured in multiples of the length of the speech signal, the real time factor xRT. For a speech signal of 2h, a processing time of 2xRT corresponds to 4h of calculation.

more difficult to obtain for resampling techniques. Therefore, a fixed size output layer was used and the words in the shortlist were rearranged in order to occupy contiguous locations in memory.

In our initial implementation we used standard stochastic backpropagation and double precision for the floating point operations in order to ensure good convergence. Despite careful coding and optimized BLAS libraries [Intel's MKL, 2004; ATLAS, 2004] for the matrix/vector operations, one epoch through a training corpus of 12.4M examples took about 47 hours on a Pentium Xeon 2.8 GHz processor. This time was reduced by a factor of more than 30 using the following techniques [Schwenk, 2004]:

- **Floating point precision** (1.5 times faster). Only a slight decrease in performance was observed due to the lower precision.
- **Suppression of intermediate calculations** when updating the weights (1.3 times faster).
- **Bunch mode**: forward and back-propagation of several examples at once (up to 10 times faster).
- **Multi-processing**: use of SMP-capable BLAS libraries for off-the-shelf bi-processor machines (1.5 times faster).

Most of the improvement was obtained by using bunch mode in the forward and backward pass. After calculating the derivatives of the error function ΔB at the output layer, the following equations were used (similar to [Bilmes *et al.*, 1997]):

$$b = b - \lambda \Delta Bi \qquad (6.16)$$

$$\Delta D = U^T \Delta B \qquad (6.17)$$

$$U = -\lambda \Delta B Z^T + \alpha U \qquad (6.18)$$

$$\Delta D = \Delta D \odot (1 - Z \odot Z) \qquad (6.19)$$

$$d = d - \lambda \Delta Di \qquad (6.20)$$

$$\Delta X = H^T \Delta D \qquad (6.21)$$

$$H = -\lambda \Delta D X^T + \alpha H \qquad (6.22)$$

where $i = (1, 1, \ldots 1)^T$, with dimension of the bunch size. The symbol $\odot$ denotes element-wise multiplication. Note that the backpropagation and weight update step, including weight decay, is done in one operation using the GEMM function of the BLAS library (Eq. 6.18 and 6.22). For this, the weight decay factor ε is incorporated into $\alpha = 1 - \lambda\varepsilon$. The update step of the projection matrix is not shown, for clarity.

Table 6.7 summarizes the effect of the different techniques to speed up training. Extensive experiments were first done with a training corpus of 1.1M

Size of Training data	Double Prec.	Float Prec.	Bunch mode						SMP 128
			2	4	8	16	32	128	
1.1M words	2h	1h16	37m	31m	24m	14m	11m	8m18	5m50
12.4M words	47h	30h	10h12	8h18	6h51	4h01	2h51	2h09	1h27

Table 6.7 Training times (for one epoch) reflecting the different improvements (on a Intel Pentium CPU at 2.8 GHz)

words and then applied to a larger corpus of 12.4M words. Bilmes et al. reported that the number of epochs needed to achieve the same MSE increases with the bunch size [Bilmes *et al.*, 1997]. In our experiments the convergence behavior also changed with the bunch size, but after adapting the learning parameters of the neural network only small losses in perplexity were observed, and there was no impact on the word error when the neural LM was used in lattice rescoring.

6.7.3 Regrouping of training examples

The above described bunch mode optimization is generic and can be applied to any neural network learning problem, although it is most useful for large tasks like this one. In addition we propose new techniques that rely on the particular characteristics of the language model training corpus. A straightforward implementation of stochastic backpropagation is to cycle through the training corpus, in random order, and to perform a forward/backward pass and weight update for each 4-gram. However, in large texts it is frequent to encounter some 4-grams several times. This means that identical examples are trained several times. This is different from other pattern recognition tasks, for instance optical character recognition, for which it is unlikely to encounter twice the exact same example since the inputs are usually floating point numbers. In addition, for the LM task, we will find many occurrences of the same context in the training corpus for which several different target words should be predicted. In other words, we can say that for each trigram there are usually several corresponding 4-grams. This fact can be used to substantially decrease the number of operations. The idea is to regroup these examples and to perform only one forward and backward pass though the neural network. The only difference is that there are now multiple output targets for each input context. Furthermore, 4-grams appearing multiple times can be learned at once by multiplying the corresponding gradients by the number of occurrences. This is equivalent to using bunch mode where each bunch includes all examples with the same trigram context. Alternatively, the cross-entropy targets can also be set to the empirical posterior probabilities (rather than 0 or 1), i.e. the relative frequencies of the 4-grams with a common context.

Words in corpus	229k	1.1M	12.4M
Random presentation:			
# 4-grams	162k	927k	11.0M
Training time	144s	11m	171m
Regrouping:			
# distinct 3-grams	124k	507k	3.3M
Training time	106s	6m35s	58m

Table 6.8 Training times for stochastic backpropagation using random presentation of all 4-grams in comparison to regrouping all 4-grams that have a common trigram context (bunch size=32).

In stochastic backpropagation with random presentation order the number of forward and backward passes corresponds to the total number of 4-grams in the training corpus which is roughly equal to the number of words[10]. With the new algorithm the number of forward and backward passes is equal to the number of *distinct trigrams* in the training corpora. One major advantage of this approach is that the expected gain, i.e. the relation between the total number of 4-grams and distinct trigrams, increases with the size of the training corpus. Although this approach is particularly interesting for large training corpora (see Table 6.8), we were not able to achieve the same convergence as with random presentation of individual 4-grams. Overall the perplexities obtained with the regrouping algorithm were slightly higher.

6.7.4 Baseline Speech Recognizer

The above described neural network language model is evaluated in a state-of-the-art speech recognizer for conversational telephone speech

---

[10]The sentence were surrounded by begin and end of sentence markers. The first two words of a sentence do not form a full 4-gram, i.e. a sentence of 3 words has only two 4-grams.

[Gauvain *et al.*, 2003]. This task is known to be significantly more difficult than the recognition of Wall Street journal or of broadcast news data. Based on the NIST speech recognition benchmarks [Lee *et al.*, 2003; Fiscus *et al.*, 2004], current best broadcast news transcription systems achieve word error rates that approach 10% in 10xRT (10 times real time) while the word error rate for the DARPA conversational telephone speech recognition task is about 20% using more computational resources (20xRT). A large amount of this difference can of course be attributed to the difficulties in acoustic modeling, but language modeling of conversational speech also faces problems that are much less important in broadcast news data such as unconstrained speaking style, hesitations, etc.

The word recognizer uses continuous density hidden Markov models (HMM) with Gaussian mixtures for acoustic modeling and n-gram statistics estimated on large text corpora for language modeling. Each context-dependent phone model is a tied-state left-to-right context dependant HMM with Gaussian mixture observation densities where the tied states are obtained by means of a decision tree. The acoustic feature vector has 39 components including 12 cepstrum coefficients and the log energy, along with their first and second derivatives. Cepstral mean and variance normalization are carried out on each conversation side. The acoustic models are trained using the maximum mutual information criterion on up to 2200 hours of data.

Decoding is carried out in three passes. In the first pass the speaker gender for each conversation side is identified using Gaussian mixture models, and a fast trigram decode is performed to generate approximate transcriptions. These transcriptions are used to compute the vocal tract length normalization (VTLN) warp factors for each conversation side and to adapt the speaker adaptive (SAT) models that are used in the second pass. Passes 2 and 3 make use of the VTLN-warped data to generate a trigram lattice per speaker turn which is expanded with the 4-gram baseline back-off LM and converted into a confusion network with posterior probabilities. The best hypothesis in the confusion network is used in the next decoding pass for unsupervised maximum likelihood linear regression (MLLR) adaptation of the acoustic models (constraint and unconstrained). The third pass is similar to the second one but more phonemic regression classes are used and the search space is limited to the word graph obtained in the second pass. The overall run time is about 19xRT.

6.7.5 Language model training

The main source for training a language model for conversational speech are the transcriptions of the audio training corpora. Three different *in-domain* data corpora have been used:

- **7.2M words:** Our first experiments were carried out with the initial release of transcriptions of acoustic training data for the Switchboard (SWB) task, namely the *careful* transcriptions of the SWB corpus distributed by LDC (2.7M words) and by the Mississippi State University (ISIP) (2.9M words), the Callhome corpus (217k words), some SWB cellular data (230k words) and *fast* transcriptions of a previously unused part of the SWB2 corpus (80h, 1.1M word).
- **12.3M words:** In 2003 LDC changed the data collection paradigm of conversational data to the so called "Fisher-protocol". Fast transcriptions of 520h of such data were available for this work.
- **27.2M words:** In a second release, additional fast transcriptions have been made available, doubling the total amount of language model training data.

All this data is available from the linguistic data consortium (LDC).[11] In addition to these in-domain corpora the following texts have been used to train the 4-gram back-off LM:

- 240M words of commercially produced broadcast news transcripts,
- 80M words of CNN television broadcast news transcriptions,
- up to 500M words of conversational like data that was collected from the Internet[12].

We refer to these data as the broadcast news corpus. Adding other sources, in particular newspaper text, did not turn out to be useful. The LM vocabulary contains 51k words and the fraction of test words not in the training set is 0.2%. The baseline LM is constructed as follows. Separate back-off n-gram language models are estimated on all the above corpora using the modified version of Kneser-Ney smoothing as implemented in

---

[11]http://www.ldc.upenn.edu/
[12]This data has been provided by the University of Washington.

the SRI LM toolkit [Stolcke, 2002]. A single back-off LM was built by merging these models, estimating the interpolation coefficients with an EM procedure. Table 6.9 gives some statistics about the reference language models.

In-domain data [words] (+broadcast news corpus)	7.2M	12.3M	27.2M
Number of 2-grams	20.1M	20.1M	26.1M
3-grams	38.8M	31.5M	40.3M
4-grams	24.0M	24.3M	39.4M
Perplexity on Eval03 test	53.0	51.5	47.5

Table 6.9 Statistics for the back-off 4-gram reference LM built using in-domain training data of varying sizes and more than 500M words of broadcast news data.

6.7.6 Experimental results

The neural network LM was trained only on the in-domain corpora (7 M, 12M and 27M words respectively). Two experiments have been conducted:

1. The neural network LM is interpolated with a back-off LM that was trained only on the in-domain copora and compared to this LM,
2. The neural network LM is interpolated with the full back-off LM (in-domain and broadcast news data) and compared to this full LM.

The first experiment allows us to assess the real benefit of the neural LM since the two smoothing approaches (back-off and neural network) are compared on the same data. In the second experiment all the available data are used to obtain the overall best results. The perplexities of the neural network and the back-off LM are given in Table 6.10.

A perplexity reduction of about 9% relative is obtained independently of the size of the LM training data. This gain decreases to approximately 6% after interpolation with the back-off LM trained on the additional broadcast news corpus of out-of domain data. It can been seen that the perplexity of the neural network LM trained only on the in-domain data is better than that of the back-off reference LM trained on all data (45.5 with respect to 47.5). Despite these rather small gains in perplexity, consistent word error reductions were observed (see Figure 6.4). The first system is that described in [Gauvain *et al.*, 2003].

In-domain data [words]	7.2M	12.3M	27.2M
In-domain data only:			
Back-off LM	62.4	55.9	50.1
Neural LM	57.0	50.6	45.5
Interpolated with all data:			
Back-off LM	53.0	51.1	47.5
Neural LM	50.8	48.0	44.2

Table 6.10 Eval03 test set perplexities for the back-off and neural LM as a function of the size of the in-domain training data.

The second system has a much lower word error rate than the first one due to several improvements of the acoustics models, and the availability of more acoustic training data. The third system differs from the second one again by the amount of training data used for the acoustic and the language model.

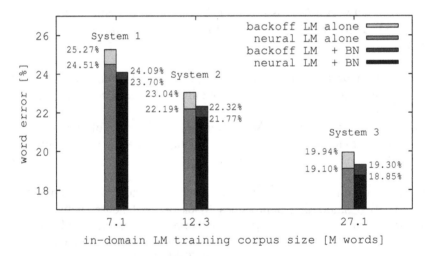

Fig 6.5 Word error rates on the Eval03 test set for the back-off LM and the neural network LM, trained only on in-domain data (left bars for each system) and interpolated with the broadcast news LM (right bars for each system).

Although the size of the LM training data has almost quadrupled from 7.2M to 27.2M words, a consistent absolute word error reduction of 0.55% can be observed [Schwenk and Gauvain, 2004]. In all these experiments, it seems that the word error reductions brought by the neural network LM are independent of the other improvements, in particular those obtained by better acoustic modeling and by adding more language model training data. When only in-domain data is used (left bars for each system in Figure 6.4) the neural network LM achieves an absolute word error reduction of about 0.8%. Note also that the neural network LM trained on at least 12.3M words is better than the back-off LM that uses in addition more than 500M words of the broadcast news corpus (system 2: 22.19 with respect to 22.32%, and system 3: 19.10 with respect to 19.30%). In a contrastive experiment, back-off and neural LMs were trained on a random 400k word subset of the 27.2M corpus, simulating for instance a domain specific task with very little LM training data. In this case, the neural network LM decreased the perplexity from 84.9 to 75.6 and the word error rate from 25.60% to 24.67%.

6.7.7 Interpolation and ensembles

When building back-off LM for a collection of training corpora, better results are usually obtained by first building separate language models for each corpus and then merging them together using interpolation coefficients obtained by optimizing the perplexity on some development data [Jelinek and Mercer, 2000]. This procedure has been used for all our back-off language models. It is, however, not so straightforward to apply this technique to a neural network LM since individually trained language models can not be merged together, and they need to be interpolated on the fly. In addition, it may be sub-optimal to train the neural network language models on subparts of the training corpus since the continuous word representations are not learned on all data. For this reason, in the above described experiments only one neural network was trained on all data.

Training large neural networks (2M parameters) on a lot of data (27.2M examples) with stochastic backpropagation may pose convergence problems and we believe that the neural network underfits the training data. Unfortunately, more sophisticated techniques than stochastic gradient descent are not tractable for problems of this size. On the other hand, several well known ensemble learning algorithms used in the machine learning community can be applied to neural networks, in particular Bagging and AdaBoost. Following to the bias/variance decomposition of error, Bagging improves performance by minimizing the variance of the classifiers [Breiman, 1994], while AdaBoost sequentially constructs classifiers that try to correct the errors of the previous ones [Freund, 1995]. In this work we have evaluated another variance reduction method: several networks are trained on all the data, but after each epoch the training data is randomly shuffled.

# hidden units	1024	1280	1560	1600	2000
One network	22.19	22.22	22.18	-	-
Ensembles	2x500	3x400	3x500	4x400	4x500
	22.15	22.12	22.07	22.09	22.09

Table 6.11 Word error rates for one large neural network and ensembles, trained on the 12.4M word in-domain corpus only.

As can been seen in Table 6.11 increasing the number of parameters of one large neural network does not lead to improvements for more than 1000 hidden units, but using ensembles of several neural networks with the same total number of parameters results in a slight decrease in the word error rate. As a side effect, training is also faster since the individual smaller networks can be trained in parallel on several machines, without any communication overhead.

6.7.8 Evaluation on other data

The neural network language model was also evaluated on other languages and tasks, in particular English and French broadcast news (BN) and English and Spanish parliament speeches (PS) [Schwenk and Gauvain, 2005]. In all cases it achieved significant word error reductions with respect to the reference back-off language model (see Table 6.12) although it was trained on less data. For each task the amount of available language model training data, the word error rate and the processing time are given.

Task	Back-off LM			Neural network LM		
	LM data	Werr	Runtime	LM data	Werr	Addtl. Runtime
French BN	523M	10.7%	8xRT	22M	10.4%	0.30xRT
English PS	325M	11.5%	10xRT	34M	10.8%	0.10xRT
Spanish PS	36M	10.7%	10xRT	34M	10.0%	0.07xRt

Table 6.12 Evaluation of the neural network language model in a speech recognizer for broadcast news (BN) and Parliament speeches (PS).

6.8 Conclusions and Future Work

This chapter has introduced an approach to learn a distributed representation for language models using artificial neural networks, a number of techniques to speed up training and testing of the model, and successful applications of the model, notably for speech recognition.

The basic idea behind this model is that of learning an embedding for words that allows to generalize to new sentences composed of words that have some similarity to words in training sentences. The embedding is continuous and the model learns a smooth map from the embeddings of the words in the sentence to the conditional probability of the last word. Hence when a word is replaced by a similar one, a similar conditional probability function is computed. Instead of distributing probability mass according to exact match of subsequences, as in n-gram models, a more subtle notion of similarity in the space of word sequences is learnt and applied.

One of the difficulty with this approach, however, compared to n-gram models, is that it requires a large number of multiply-adds for each prediction, whereas n-gram models only require a few table look-ups. A good deal of the chapter therefore explores different methods to speed-up the algorithms. The computational bottleneck involves the loop or the sum over the set of words over which the conditional probability is to be computed. The three different methods all aim at reducing this computation, either by sampling, by hierar-chical decomposition of the probabilities, or by focusing on the most frequent words. Speed-up on the order of 100-fold are thus obtained and a large-scale application of the approach on a state-of-the-art speech recognition system shows that it can be used to improve on the state-of-the-art performance.

Acknowledgments

The authors would like to thank Pascal Vincent, Réjean Ducharme and Christian Jauvin for their collaboration on the neural language model, and Léon Bottou, Yann Le Cun and Geoffrey Hinton for useful discussions. The authors would also like to recognize the contributions of Lori Lamel, Gilles Adda and Fabrice Lefèvre for their involvement in the development of the LIMSI conversational speech recognition system from which part of this work is based. Part of this research was made possible by funding to Yoshua Bengio from the NSERC granting agency, as well as the MITACS and IRIS networks.

References

Automatically tuned linear algebra software, https://sourceforge.net/projects/math-atlas/atlas

Baker, D. and McCallum, A. (1998). Distributional clustering of words for text classification. In SIGIR'98.

Bellegarda, J. (1997). A latent semantic analysis framework for large–span language modeling. In Proceedings of Eurospeech 97, pages 1451–1454, Rhodes, Greece.

Bengio, S. and Bengio, Y. (2000a). Taking on the curse of dimensionality in joint distributions using neural networks. IEEE Transactions on Neural Networks, special issue on Data Mining and Knowledge Discovery, 11(3), 550–557.

Bengio, Y. and Bengio, S. (2000b). Modeling high-dimensional discrete data with multi-layer neural networks. In S. Solla, T. Leen, and K.-R. Muller, editors, Advances in Neural Information Processing Systems 12, pages 400–406. MIT Press.

Bengio, Y. and Senecal, J.-S. (2003). Quick training of probabilistic neural nets by sampling. In Proceedings of the Ninth International Workshop on Artificial Intelligence and Statistics, volume 9, Key West, Florida. AI and Statistics. 38 Authors Suppressed Due to Excessive Length

Bengio, Y., Ducharme, R., Vincent, P., and Jauvin, C. (2003). A neural probabilistic language model. Journal of Machine Learning Research, 3, 1137–1155.

Berger, A., Della Pietra, S., and Della Pietra, V. (1996). A maximum entropy approach to natural language processing. Computational Linguistics, 22, 39–71.

Bilmes, J., Asanovic, K., Chin, C.-W., and Demmel, J. (1997). Using phipac to speed error back-propagation learning. In International Conference on Acoustics, Speech, and Signal Processing, pages V: 4153–4156.

Blitzer, J., K.Q.Weinberger, Saul, L., and Pereira, F. (2005). Hierarchical distributed representations for statistical language models. In L. Saul, Y. Weiss, and L. Bottou, editors, Advances in Neural Information Processing Systems 17. MIT Press.

Breiman, L. (1994). Bagging predictors. Machine Learning, 24(2), 123–140.

Brown, A. and Hinton, G. (2000). Products of hidden markov models. Technical Report GCNU TR 2000-004, Gatsby Unit, University College London.

Brown, P., Pietra, V. D., DeSouza, P., Lai, J., and Mercer, R. (1992). Class-based n-gram models of natural language. Computational Linguistics, 18, 467–479.

Chen, S. F. and Goodman, J. T. (1999). An empirical study of smoothing techniques for language modeling. Computer, Speech and Language, 13(4), 359–393.

Cheng, J. and Druzdzel, M. J. (2000). Ais-bn: An adaptive importance sampling algorithm for evidential reasoning in large Bayesian networks. Journal of Artificial Intelligence Research, 13, 155–188.

Deerwester, S., Dumais, S., Furnas, G., Landauer, T., and Harshman, R. (1990). Indexing by latent semantic analysis. Journal of the American Society for Information Science, 41(6), 391–407.

Elman, J. (1990). Finding structure in time. Cognitive Science, 14, 179–211.

Emami, A., Xu, P., and Jelinek, F. (2003). Using a connectionist model in a syntactical based language model. In International Conference on Acoustics, Speech, and Signal Processing, pages I: 272–375.

Fellbaum, C. (1998). WordNet: An Electronic Lexical Database. MIT Press.

Fiscus, J., Garofolo, J., Lee, A., Martin, A., Pallett, D., Przybocki, M., and Sanders, G. (Nov 2004). Results of the fall 2004 STT and MDE evaluation. In DARPA Rich Transcription Workshop, Palisades NY.

Freund, Y. (1995). Boosting a weak learning algorithm by majority. Information and Computation, 121(2), 256–285.

Gauvain, J.-L., Lamel, L., Schwenk, H., Adda, G., Chen, L., and Lefevre, F. (2003). Conversational telephone speech recognition. In International Conference on Acoustics, Speech, and Signal Processing, pages I: 212–215.

Goodman, J. (2001a). A bit of progress in language modeling. Technical Report MSR-TR-2001-72, Microsoft Research.

Goodman, J. (2001b). Classes for fast maximum entropy training. In International Conference on Acoustics, Speech, and Signal Processing, Utah.

Hinton, G. (1986). Learning distributed representations of concepts. In Proceedings of the Eighth Annual Conference of the Cognitive Science Society, pages 1–12, Amherst 1986. Lawrence Erlbaum, Hillsdale.

Hinton, G. (2000). Training products of experts by minimizing contrastive divergence. Technical Report GCNU TR 2000-004, Gatsby Unit, University College London.

Hinton, G. (2002). Training products of experts by minimizing contrastive divergence. Neural Computation, 14(8), 1771–1800. 1 Neural Probabilistic Language Models 39

Hinton, G. and Roweis, S. (2003). Stochastic neighbor embedding. In S. Becker, S. Thrun, and K. Obermayer, editors, Advances in Neural Information Processing Systems 15. MIT Press, Cambridge, MA.

Jelinek, F. and Mercer, R. (2000). Interpolated estimation of markov source parameters from sparse data. Pattern Recognition in Practice, pages 381–397.

Jelinek, F. and Mercer, R. L. (1980). Interpolated estimation of Markov source parameters from sparse data. In E. S. Gelsema and L. N. Kanal, editors, Pattern Recognition in Practice. North-Holland, Amsterdam.

Jensen, K. and Riis, S. (2000). Self-organizing letter code-book for text-to-phoneme neural network model. In Proceedings ICSLP.

Katz, S. M. (1987). Estimation of probabilities from sparse data for the language model component of a speech recognizer. IEEE Transactions on Acoustics, Speech, and Signal Processing, ASSP-35(3), 400–401.

Kneser, R. and Ney, H. (1995). Improved backing-off for m-gram language modeling. In International Conference on Acoustics, Speech, and Signal Processing, pages 181–184.

Kong, A. (1992). A note on importance sampling using standardized weights. Technical Report 348, Department of Statistics, University of Chicago.

Kong, A., Liu, J. S., and Wong, W. H. (1994). Sequential imputations and Bayesian missing data problems. Journal of the American Statistical Association, 89, 278–288.

Lee, A., Fiscus, J., Garofolo, J., Przybocki, M., Martin, A., Sanders, G., and Pallett, D. (May 2003). Spring speech-to-text transcription evaluation results. In Rich Transcription Workshop, Boston.

Liu, J. S. (2001). Monte Carlo Strategies in Scientific Computing. Springer.

Luis, O. and Leslie, K. (2000). Adaptive importance sampling for estimation in structured domains. In Proceedings of the 16th Annual Conference on Uncertainty in Artificial Intelligence (UAI-00), pages 446–454.

Intel math kernel library (2004)., http://www.intel.com/software/products/mkl/.

Miikkulainen, R. and Dyer, M. (1991). Natural language processing with modular neural networks and distributed lexicon. Cognitive Science, 15, 343–399.

Ney, H. and Kneser, R. (1993). Improved clustering techniques for class-based statistical language modeling. In European Conference on Speech Communication and Technology (Eurospeech), pages 973–976, Berlin.

Niesler, T., Whittaker, E., and Woodland, P. (1998). Comparison of part-of-speech and automatically derived category-based language models for speech recognition. In International Conference on Acoustics, Speech, and Signal Processing, pages 177–180.

Paccanaro, A. and Hinton, G. (2000). Extracting distributed representations of concepts and relations from positive and negative propositions. In Proceedings of the International Joint Conference on Neural Network, IJCNN'2000, Como, Italy. IEEE, New York.

Pereira, F., Tishby, N., and Lee, L. (1993). Distributional clustering of English words. In 30th Annual Meeting of the Association for Computational Linguistics, pages 183–190, Columbus, Ohio.

Riis, S. and Krogh, A. (1996). Improving protein secondary structure prediction using structured neural networks and multiple sequence profiles. Journal of Computational Biology, pages 163–183. 40 Authors Suppressed Due to Excessive Length

Robert, C. P. and Casella, G. (2000). Monte Carlo Statistical Methods. Springer. Springer texts in statistics.

Salton, G. and Buckley, C. (1988). Term weighting approaches in automatic text retrieval. Information Processing and Management, 24(5), 513–523.

Schmidhuber, J. (1996). Sequential neural text compression. IEEE Transactions on Neural Networks, 7(1), 142–146.

Schutze, H. (1993). Word space. In C. Giles, S. Hanson, and J. Cowan, editors, Advances in Neural Information Processing Systems 5, pages pp. 895–902, San Mateo CA. Morgan Kaufmann.

Schwenk, H. and Gauvain, J.-L. (2002). Connectionist language modeling for large vocabulary continuous speech recognition. In International Conference on Acoustics, Speech, and Signal Processing, pages I: 765–768.

Schwenk, H. and Gauvain, J.-L. (2003). Using continuous space language models for conversational speech recognition. In ISCA & IEEE Workshop on Spontaneous Speech Processing and Recognition, Tokyo.

Schwenk, H. (2004). Efficient training of large neural networks for language modeling. In IEEE joint conference on neural networks, pages 3059–3062.

Schwenk, H. and Gauvain, J.-L. (2004). Neural network language models for conversational speech recognition. In International Conference on Speech and Language Processing, pages 1215–1218.

Schwenk, H. and Gauvain, J.-L. (2005). Building Continuous Language Models for Transcribing European Languages. In Eurospeech. To appear.

Stolcke, A. (2002). SRILM - an extensible language modeling toolkit. In Proceedings of the International Conference on Statistical Language Processing, Denver, Colorado.

Xu, W. and Rudnicky, A. (2000). Can artificial neural network learn language models? In International Conference on Statistical Language Processing, pages M1–13, Beijing, China.

7 Computational Grammatical Inference

Pieter W. Adriaans[1], Menno M. van Zaanen[2]

[1]ILLC, University of Amsterdam, Amsterdam, the Netherlands,
pietera@science.uva.nl
[2]ILK, Tilburg University, Tilburg, the Netherlands, mvzaanen@uvt.nl

Abstract

Grammatical Inference (GI) concentrates on finding compact representations, i.e. grammars, of possibly infinite sets of sentences. These grammars describe what sentences do or do not belong to a particular language. The process of learning the form of a grammar based on example sentences from the language touches several fields. Here, we give an overview of the field of GI as well as fields that are closely related. We discuss linguistic, empirical, and formal grammatical inference and discuss the work that falls in the areas where these fields overlap.

7.1 Introduction

A grammar is a finite structure that describes a possibly infinite set of sentences, called a language. In this chapter we are concerned with learning grammars from unstructured data. The unstructured data is a sample taken from the language the grammar describes. This process is called Grammatical Inference (GI) and can be researched from many different angles. Here we will relate the linguistic, empirical and formal aspects of GI with respect to the field of computational grammatical inference (CGI).

CGI is mainly concerned with finding computational models that can learn grammars. The models can have linguistic impact, but can also serve as a practical tool to experiment with the theoretical proofs taken from the field of formal GI.

In this chapter we discuss the different fields related to the research performed in the field of CGI. We do this starting from the "applied" GI in linguistics and end in the "theoretical" field of formal GI, via CGI. Of course, the boundaries between the research areas are not always so strict and clear cut. Many results are applicable to more than one field, which is really why these research topics should be considered in total. During our

P.W. Adriaans and M.M. van Zaanen: *Computational Grammatical Inference*, StudFuzz **194**, 187–203 (2006)
www.springerlink.com   © Springer-Verlag Berlin Heidelberg 2006

discussion, you will find that, unfortunately, there has not been much interaction between researchers from the different areas. This will also be our final conclusion.

7.2 Linguistic Grammatical Inference

Within linguistics, grammatical inference has an impact in different fields. Here, we will focus mainly on first language acquisition, but much of the discussion that follows also applies to other areas.

To get an idea of the extend of the problem of grammatical inference, we mention the discussion about the complexity of human languages. The general consensus is that natural languages are at least context-free and with high probability context-sensitive. (Huybrechts, 1984; Shieber, 1985) We will mainly discuss learning context-free languages as a first approximation of human languages. So far, there has been hardly any research into the learnability of context-sensitive languages. It is unlikely that the full expressive power of the class of context-sensitive languages is necessary to describe a natural language, however, context-sensitive constructions seem to exist in some cases.

Research on language learning has led to major controversies within the field of linguistics. One very influential reason for this is the famous conjecture of Chomsky that the efficiency of human language acquisition can only be explained on the basis of an innate Universal Grammar (UG). The reason for this would be that the learner is not provided with enough linguistic information to learn the linguistic structures as fast as it does. This is called "the poverty of the stimulus". The UG discussion is in fact is a revival of the ancient philosophical debate of rationalism (Descartes) versus empiricism (Hume). This argument has attracted a lot of attention, unfortunately with surprisingly little tangible results.

Past research in this field has shown that a proper study of language acquisition will require a huge interdisciplinary enterprise. Not only is there the need for theoretical research, but also a huge empirical effort has to be made. This requires co-operation between linguists, psychologists, audiologists, neuro-physiologists, cognitive scientists, computational linguists, and computer scientists.

To get an idea of the impact of the amount of work required, imagine the creation of elementary annotated video sequences to study the language development of children. This is an investment that easily amounts to several hundreds, if not thousands of person-years. For a statistically significant experiment one needs a group of at least some twenty children (a target group of ten, a control group of ten). Then take one hour of video per child per week for the duration of three years and annotate the behavioral and dialogical interaction between mother and child, the phonetic and orthographic transcription, and include further linguistic analysis. This is already takes roughly 150 person-years. Now, multiply this by a number of different languages and a number of different target groups (hearing impaired children, bi-lingual children, etc.) one would like to analyze. Also, one would need to add the costs of ERP and fMRI scans. To put this huge project into proportions, observe that some experts think that one hour of video per week is insufficient to capture the sudden increase in linguistic competence that children display in a very limited time span.

It may be clear that a proper understanding of natural language acquisition is not simply a matter of choice between philosophical schemes like tabula rasa versus innate ideas. The only option we have now is to postpone the difficult debate until more empirical evidence is available. At the moment, we should focus on smaller sub-problems in language acquisition, keeping the complete problem in mind. Current research concentrates on a thorough understanding of the various stages of language acquisition and tries to understand the complexity of transformations between different stages.

Not only the means of acquiring language is debated, the efficiency of human language acquisition is also ill-understood at this moment. The explanation Chomsky provided, i.e. the influence of universal linguistic categories, has induced controversy. There is hardly any agreement on general principles and even the universality of the distinction between nouns and verbs has been doubted.[1]

A related, and perhaps as important, aspect in this context is the distinction between open and closed classes of words. Open classes can be extended without changing the inherent structure of the language (examples of such classes are nouns, verbs, or adjectives), whereas closed classes are functional categories such as prepositions, determiners and auxiliaries which contain about a dozen words each. Trying to change these classes would result in a thorough change of the actual language. An additional difference between the words in the classes is that open class words can be paraphrased or described. This is not possible for words from the closed class. (Try to describe the word "the" or "in" in other words.) Even though this seems to be a universal aspect of language, actual languages seem to vary considerably in terms functional categories. In some languages the role of the functional categories is partly taken over by morphology (e.g. Latin).

---

[1] See http://linguistics.arizona.edu/~carnie/papers/V1Volume/Gil.pdf.

One of the main problems of language learning seems to be the acquisition of the functional classes. Where the open classes require unlimited bootstrapping (or at least induction) capability, the words in the closed classes cannot be learned as easily. Currently, a working hypothesis that the closed classes are induced based on samples taken from limited sets of open classes is used. Initial studies by (Pinker, 1999) (semantic bootstrapping) and (van Kampen, 1997) (syntactic bootstrapping) look promising, but more empirical research is needed.

Note also that human language learning is extremely flexible. There is no evidence that phases of linguistic development are universal for all languages or even that subjects that learn the same language will take the same learning route. The actual flexibility can be observed when looking at, for example, hearing impaired subjects. The input signal is different, some clues that might be useful for language learning are missed. This still allows for language learning (albeit perhaps in different forms, such as sign language). This indicates that different, non-standard learning routes have been taken, showing the flexibility in the process. A better insight in these non-standard learning paths will help in better understanding the underlying processes.

A different approach to investigate language learning can be found in the field of cognitive neuro-science. Researchers have identified interesting ERP phenomena such as the N400 effect that is associated with semantic processing and the P600/SPS effect that has a relation with the syntactic analysis in the human brain. This has led to various neuro-computational models that have recently been proposed. (Hagoort, 2003), for example, presents a neuro-computational model of syntactic processing based on a computational model of parsing due to (Vosse,2000). The main impact of these proposals is directed towards the binding problem. The grammatical model underlying the binding solution must be learned and this work researches development in the brain.

7.3 Empirical Grammatical Inference

Whereas linguistic grammatical inference is concerned with understanding language learning from a linguistic point of view, the field of empirical grammatical inference (also called Computational Grammar Inference (CGI)) deals with the creation of computational models for identification of language, based on a finite set of examples. CGI is a research field in its own right, with its own research questions, that do not necessarily have direct impact on the linguistic study of human language learning. However, the computational models developed in this field are of interest to the linguistic community. The computational models try to learn a grammar in relatively short time, a task similar to the language learning of humans.

By building computational language learning models, we hope to get a better understanding of the underlying bias that governs the probability distribution of human communication. One of the research questions is whether this universal bias might be able to solve the problem of poverty of stimulus. Alternatively, the bias might explain why human language contains certain specific syntactic constructions.

Within CGI, several approaches are considered. Firstly, an initial estimate of the bias can be made by investigating the relevant probability distributions. This can be done using algorithmic complexity theory, one of the major achievements in computer science in the previous century. The idea here is to describe the relationship between the notion of computational complexity of an object and it's a priori probability. The basic insight here is that objects that are easier to compute have a higher probability. With respect to grammatical inference, this results in a first approximation of co-operative linguistic behavior: if one wants to teach a language, it appears reasonable to select examples that are easy to analyze first. Using algorithmic complexity theory one can actually give adequate answers to questions like: Given a text T and two grammars G_1 and G_2 are we able to approximate formula 1?

$$\arg\max\{P(G_1 \mid T), P(G_2 \mid T)\} \tag{1}$$

Secondly, a group of researchers develop grammatical inference algorithms that actually try to infer structure using language data. These algorithms can be implemented and tested using natural language collections. The results can be measured, showing effectiveness of the algorithms. An example of this approach can be found in (Wolff, 2003) which describes the ICMAUS framework, based on information compression and ideas of (Solomonoff, 1997) and Rissanen. Another system comes from (Adriaans, 1992) who developed the EMILE framework, which has culminated in the EMILE 4.1 implementation by (Adriaans, 2002). A methodology related to the ICMAUS framework, is presented by (van Zaanen, 2002). This framework is called Alignment-Based Learning.

This empirical approach to CGI has some specific problems. One of these is the means of evaluation. (van Zaanen, 2001) compare the results generated by two CGI implementations by applying them to two treebanks. Even though this manner of evaluation has developed into a de facto standard, (van Zaanen, 2004) described several problems with all existing approaches. A new approach where systems are applied to treebanks in several languages and annotated using different annotations schemes is being developed.

Whereas the GI methods described above are mainly concerned with learning context-free grammars, there is also much research performed on the efficient learning of regular languages. This research takes place mainly in the ICGI community.[2] This has lead to algorithms that efficiently learn complex Deterministic Finite Automata (DFA) with several hundreds of internal states. A good example of state of the art is the so-called blue fringe algorithm. (Lang, 1998)

7.4 Formal Grammatical Inference

In the field of formal grammatical inference, people research the identification of finite descriptions or structures that describe a possibly infinite set of sentences based on a finite number of examples. Hume (1909) in the 17th century noticed that such form of induction cannot be made deductively valid. In general, it is impossible to derive universal rules on the basis of a finite set of examples. This principle can be found described in term so grammatical inference in the work of Gold.

In (Gold, 1967), a game theoretic definition of learnability is proposed. In so called *identification in the limit*, there are two parties, the Challenger and the Learner. The task of the Learner is to try and identify the language, the Challenger is trying to communicate. First of all, a class of possible grammars is selected. The Challenger selects one of these grammars and presents an enumeration of structures to the Learner.

When the Challenger provides plain sentences taken from the language, the process is called learning from *positive information*. However, if negative information is also provided as such, i.e. sentences that are not in the language, we speak of *complete information*.

The Learner produces an infinite number of guesses. A language is learnable in the limit when the guesses of the Learner stabilize at the right guess. It may be obvious that with respect to linguistic grammars, no interesting class of grammars is learnable from positive information only. A brief overview of some results in this framework can be found in table 1.

---

[2] International Grammar Induction Community, which organizes a two year conference on the subject as well as several grammar induction competitions: Abbadingo, Gowachin and more recently, Omphalos.

	Complete information	Positive information
Recursively Enumerable Sets	Not learnable	Not learnable
Decidable Rewriting Systems	Learnable	Not learnable
Super Finite Set	Learnable	Not learnable
Finite Sets	Learnable	Learnable
Finite Class of Finite Sets	Learnable	Learnable

Table 1. Learnability of classes of grammars.

The central theorem within the field of identification in the limit is:

Theorem [Gold]:

A class of grammars G is not learnable from positive information if G contains all finite languages and at least one infinite language that is the union of all the finite languages.

A class of language that conforms to this theorem is said to have *infinite elasticity*. This means that it has an infinite sequence of real inclusions of languages. Effectively, this means that at each point during the learning process, the Learner still has an infinite number of possible grammars from which one has to be chosen. It can be proven that only classes of languages that have *finite elasticity* can be identified in the limit.

Note that one could argue that identification in the limit is not really an inductive approach. The Learner can simply wait until enough evidence is present to deduce the actual grammar from the data given. Furthermore, there is no restriction on the structures presented by the Challenger with respect to mode of presentation. The only constraint there is, requires the Challenger to select any structure in the languages at least once in the limit.

Gold's strict identification in the limit research has shown that many interesting classes of grammars are not learnable. However, on a different note, (Horning, 1969) has shown, building on the related work by Solomonoff, that probabilistic context-free grammars are indeed learnable from positive data only. Probabilistic context-free grammars are context-free grammars where for each structure a probability is given. This idea actually models a notion of grammaticality in natural language, where almost any structure can be made grammatical in the right context. This result puts the strict unlearnability results of Gold in a new perspective.

Another way of proceeding is to extend the recursion theoretic approach of Gold. Several important results can be found in this area. Most notably, the results on tell-tale sets by (Angluin, 1980) have to be mentioned. A tell-tale set is a finite sub-set of a language, that prevents over-generalization. This work describes identification with the help of strong clues such as membership queries, where the Learner may check whether a sentence is member of the target language (Angluin, 1997), or equivalence queries, where the Learner may check whether the current hypothesis of the grammar is indeed correct and a sentence in the symmetric difference of the hypothesis and the target grammar is returned if this is not the case. It can be shown that even equivalence queries are not sufficient enough to guarantee exact identification of context-free grammars in the limit.

A related negative result by (Pitt, 1988) proves that modulo a polynomial time transformation, predicting context-free grammars is as hard as computing certain cryptographic predicates.

Overall, results on learning context-free grammars have been negative. This has led to research that focused on learning subclasses of context-free grammars. A few of these classes need to be mentioned: k-bounded context-free grammars (Angluin, 1987), (k)reversible languages (Angluin, 1982; Sakakibara, 1992) and simple deterministic languages (Yokomori, 1988; Ishiszaka, 19990). However, none of these subclasses of context-free grammars can be said to have any linguistic importance. This is less true for the work by (Kanazawa, 1995), which shows learnability of certain classes of categorial grammars.

Another approach to learnability of grammar classes is the probabilistic approach. Probabilistic context-free grammars are identifiable in the limit, as mentioned above and it is actually the only linguistically relevant class of grammars currently known to be so. However, the algorithm by Horning is too complex to be empirically usable.

Based on the algorithm of Horning, further research has been performed (and still is). (Valiant, 1984) introduced a completely new learning concept that is called "probably approximately correct" (PAC) learning. The main idea behind this approach is the notion of PAC learnability:

Definition [PAC learning]:

Let F be a concept class, $\delta : 0 \leq \delta \leq 1$ a confidence parameter, $\varepsilon : 0 \leq \varepsilon \leq 1$ an error parameter. A concept class F is *PAC learnable* if for all targets $f \in F$ and all probability distributions P on the sample space $\mathbf{U}^*$ the learning algorithm A outputs a concept $g \in F$ such that with probability $(1 - \delta)$ it holds that we have a chance on an error with $P(f \Delta g) \leq \varepsilon$ (where $f \Delta g = (f - g) \cup (g - f)$).

Instead of trying to perfectly identify a specific grammar from a class of grammars, the idea is to minimize the chance of learning an incorrect grammar without being completely sure that the right grammar has been identified. The goal is to find a grammar that is probably, approximately correct. This approach has been applied to learning Boolean concepts, but since it has been shown to be more widely applicable (Li, 1991).

Unfortunately, the pure PAC learning approach does not help us much. So far, no linguistically interesting classes of grammars have been found that are learnable within the PAC framework. This may not be so surprising, since PAC learning requires a class of grammars to be learnable under any distribution. This is obviously a very strong demand, especially with respect to natural language grammatical inference.

To make this problem even worse, we know (from empirical evidence) that probability distributions that govern natural language are heavily tailed. There is always a considerable amount of probability mass that is needed to represent unseen data. For example, the distributions of word frequencies almost always follows Zipf's law: the size of the word frequency class with frequency n is $cn^{-0.7}$, where c is a constant depending on the size of the sample. In recent years our understanding of these dimensionless distributions has increased considerably (Faloutsos, 1999), but the exact reason why they occur with such regularity in linguistic structures is ill-understood, as is the meaning of the 0.7 exponent. Unfortunately, this aspect is particularly harmful for PAC convergence.

A related approach, based on algorithmic complexity theory is PAC-s learning: PAC learning under simple distributions. Using this approach it is possible to describe the notion of a priori distribution on the strings in the language as well as a class of *simple distribution* that might serve as first approximations of the underlying distribution that governs the grammar (or with respect to natural language, that governs human communication) (Li, 1991). A significant result is the coding theorem from Levin:

Theorem [Levin]:
$$-\log \mathbf{m}(x) = -\log P_U(x) + O(1) = K(x) + O(1)$$

In this theorem, $\mathbf{m}(x)$ stands for a universal semi measure, which is a recursively enumerable semi-measure μ that recursively dominates every other enumerable semi-measure μ' i.e. $\mu(x) \geq c\mu'(x)$ for a fixed constant c independent of x. Levin showed that a universal enumerable semi-measure exists. We fix a universal semi-measure $\mathbf{m}(x)$. The semi-measure $\mathbf{m}(x)$ converges to 0 slower than any positive enumerable function which converges to 0. Of course, $\mathbf{m}(x)$ itself is not recursive. $P_U(x)$ denotes the universal a priori probability of a binary string x. This probability is defined as $P_U(x) = \sum_{U(p)=x} 2^{-|p|}$. Finally, $K(x)$ is the prefix Kolmogorov complexity of the string x. That is the length of the shortest prefix-free program that generates x on a universal Turing machine. Note that the descriptive complexity of a string x relative to a Turing machine T and a binary string y is defined as the shortest program that gives output x on input y: $K(x \mid y) = \min\{p \mid : p \in \{0,1\}^*, T(p,y) = x\}$. There is a universal Turing machine U, such that for each Turing machine T there is a constant cT, such that for all x and y, we have $K_U(x \mid y) \le K_T(x \mid y) + cT$. This definition is invariant up to a constant with respect to different universal Turing machines. Hence we fix a reference universal Turing machine U, and drop the subscript U by setting $K(x \mid y) = K_U(x \mid y)$. The Kolmogorov complexity of a binary string x can now be defined as $K(x) = K(x \mid \varepsilon)$.

Using this important theorem, (Li, 1991) showed the following completeness result.

Theorem [Learnability]:

A concept class C is learnable under $\mathbf{m}(x)$ iff C is also learnable under any arbitrary simple distribution[3]. $P(x)$ provided the samples are taken according to $\mathbf{m}(x)$.

Based on these notions, together with the notion of shallowness, (Adriaans, 2001) conjectured that natural languages are shallow. Shallowness foresees some striking properties of natural languages, such as the existence of small number word classes, of which some are quite large. This also means that they can be learned from a relatively small set of structures and the length of these structures is logarithmic in the Kolmogorov complexity of the grammar.

Definition [Kolmogorov complexity]:

A language G is called *shallow* is it has a tell-tale set $C \subseteq G$ for which $\forall s \in C : |s| \le c \log K(G)$

The usefulness of this definition with respect to learning natural languages follows from the following theorem.

---

[3] A distribution is simple if it is dominated by a recursively enumerable distribution.

Theorem [Shallowness]:
 Shallow languages can be sampled under **m** in polynomial time.

The question now is whether **m** is a good approximation of co-operative linguistic behavior. If it is the case, then the probability that a speaker selects a specific string decreases exponentially with the complexity of the string. Furthermore, if the language is shallow, it increases the probability that the language can be learned efficiently.

7.5 Overlap between the fields

Now the work in three grammatical inference sub-fields have been discussed, we will turn to the research performed at the boundaries of the fields. This research often belongs to more than one field at the same time, or the research outcomes have direct influence on research question in another field.

Formal grammatical inference has had direct influence in the nativist-empiricist discussion. The work by Gold showed that *finite elasticity* is the defining characteristic of classes of languages that are identifiable in the limit. The fact that all reasonable grammatical models of natural language have infinite elasticity combined with the general impression that children do not receive negative information from their parents[4] has served as an argument in the discussion on innateness of language in favor of a Universal Grammar. The formal proofs indicate that additional information (such as innate structures) is necessary to be able to learn the language.

Note that even though the proof by Gold has been used to advocate UG, the actual impact of the results should not be taken as strongly. Even though many important classes of languages have been shown to be unlearnable (Gold, 1967; Pitt, 1988; Angluin, 1988), there does not seem to be a clear relation to the acquisition of natural language. For example, there is an early result by (Horning, 1969), who showed that probabilistic context-free grammars are indeed identifiable in the limit.

Another aspect under discussion is the strong and restricted setup of Gold's experiments. It may be assumed that in the case of natural language learning, the parent (i.e. Challenger) is more co-operative. In this respect, the underlying probability distribution should be taken into account, which results in that the class of learnable languages is substantially larger. This also implies that more research should be directed towards finding out more about the actual underlying probability distributions reflecting sen-

---

[4] Note however, that the extent of information received by the child is also under discussion. See, for example, (Sokolov, 1994) for an overview of this discussion.

tence generation. This research is a combined effort between computational, formal and linguistic grammatical inference researchers.

Recently, the problem of innateness is addressed in a new way. Computational models of grammatical inference are applied to real language data (instead of generated data from formal grammar classes). These systems perform reasonably well on real life corpora, even though the formal aspects of these systems are not always as clear-cut. Examples of these systems can be found in (Seginer, 2003), EMILE: (Vervoort, 2000), Alignment-Based Learning: (van Zaanen, 2002), Evidence based State Merging: (Lang, 1998). Additionally, there has been much research on constraints, probability distributions and bias that make linguistically relevant classes of languages learnable (Osherson, 1997). Currently, the working hypothesis is that human language can be explained in terms of a structure that:

- facilitates efficient on-line interpretation,
- has adequate descriptive power,
- satisfies certain physical and biological constraints.

However, in general, as we have discussed before, we think it is too early to give definitive answers on this matter.

Note that research using computational models of grammatical inference have also led to other results. For example, exploration using the EMILE system has lead to the observation that current text corpora are not particularly suitable for grammatical inference. The corpora are semantically biased, for example, corpora are taken from the Internet, or the Wall Street Journal, or are transcriptions of dialogues, or simply scientific texts.

Semantic bias is a problem, since certain relations cannot be easily found using biased data. For example, Wall Street Journal texts may help in learning that *dollar* and *euro* are currencies and *executive director* and *chairman* are positions or jobs. However, it is hard to learn that all are nouns, because the distributional evidence in the texts does not support this. This is a problem with respect to grammatical inference, but semantic bias can be used to learn semantic structures, such as ontologies.

It may be clear that bootstrapping occurs on various linguistic, such as syntactic, semantic, or prosodic, levels at (roughly) the same time. See, for example,
http://www-uilots.let.uu.nl/conferences/Gala/LEARN-102-Kampen.pdf.
The actual algorithmic properties of the processes are still vague. Even though, the grammatical inference systems try to model these implicitly.

Natural language learning can be roughly grouped in several phases. A rough overview of the phases occurring in natural language learning combined with current computational models for each stage can be found in table 2.

Phase	Period	Description	Model	Learning strategy
1	0-9 months	Linking acoustics and events bab-bling	DFA	evidence based state merging, prosodic boot-strapping
2	9-24 months	Children catego-rize words into word classes and show evidence of early sensitivity to syntax word classes	Complex in-teraction between deixis and babbling	syntactic and semantic boot-strapping
3	2-3.5 years	Language meaning and syntax struc-ture is acquired, emergence of re-cursive rules	induction as first ap-proximation	Seginer, EMILE, ABL

Table 2. Phases in child language acquisition

Using insights from a combination of algorithmic information theory, complexity theory and statistics, some general aspects of human languages can already be predicted (cf. the shallowness constraint proposed by (Adriaans, 1992; Adriaans, 2001)). This has led to the hypothesis that the language learning process

- is phased,
- is hybrid, in the sense that different learning strategies might be invoked in different phases,
- is cumulative, in the sense that structures learned in phase n can be used to bootstrap structures in phase $n+1$. This does not mean that lin-guistic performance itself is growing in a monotone way. Structures that are left unanalyzed in phase n can be over-fitted in phase $n+1$.

In the past, there has been surprisingly little research that combines the different sub-fields in grammatical inference. There are various reasons for this, for example, the linguistic research in grammatical inference has mainly focused on describing properties of the learning process, whereas computational and formal grammatical inference is mainly concerned with abstract language models. We expect there to be increasing cross-fertilization between the sub-fields within grammatical inference in the near future.

7.6 Conclusion

This chapter gives an overview of the current state of grammatical inference field. We have divided the research in three sub-fields, linguistic, empirical, and formal grammatical inference, that have distinct topics and research questions. For each of the sub-fields we have discussed the main results so far and described the current questions and problems remaining.

Researchers within the several sub-fields seem to have created certain boundaries between the fields that do not allow for easy co-operation with researchers from another field. We think that these boundaries are not necessarily because of disinterest, but more because of difference in terminology and goals.

When looking at the entire field (the combination of the sub-fields), it may be clear that it is time to try to remove the (artificial) boundaries and combine the research performed within each sub-field. For example, combining results from linguistics, complexity theory, computer science and (cognitive) neuro-science might result in major breakthroughs in one or more of the separate fields.

The main research questions at this time are concerned with the understanding of human communication, underlying probability distributions, and algorithms. These questions can only be answered when natural language data and learning processes are investigated, algorithms are designed and implemented and (existing) grammar induction implementations are evaluated and researched with respect to formal learnability within several frameworks (not only the Gold, but also PAC, PAC-s and alternative paradigms).

References

Adriaans, W. P. (1992). *Language Learning from a Categorial Perspective*. PhD thesis, Universiteit van Amsterdam.

Adriaans, P. (2001). Learning shallow context-free languages under simple distributions. In Copestake, A. and Vermeulen, K., editors, *Algebras, Diagrams and Decisions in Language, Logic and Computation*. CSLI/CUP, Stanford:CA, USA.

Adriaans, P. and Vervoort, M. (2002). The EMILE 4.1 grammar induction toolbox. In Adriaans, P., Fernau, H. and van Zaanen, M., editors, *Grammatical Inference: Algorithms and Applications; 6th International Colloquium, ICGI 2002,* volume 2484 of *LNCS/LNAI,* pages 293-295. Springer-Verlag, Berlin Heidelberg, Germany.

Angluin, D. (1980). Inductive inference of formal languages from positive data. *Information and Control*, 45:117-135.

Angluin, D. (1982). Inference of reversible languages. *Journal of the Association for Computing Machinery*, 29(3):741-765.

Angluin, D. (1987). Learning k-bounded context-free grammars. Technical Report YALEU/DCS/TR-557, Yale University.

Angluin, D. (1988). Queries and concept learning. *Machine Learning*, 2:319-342.

Angluin, D., Krikis, M., Sloan, R. H., and Turán, G. (1997). Malicious omissions and errors in answers to membership queries. *Machine Learning*, 28(2-3):211-255.

Faloutsos, M., Faloutsos, P., and Faloutsos, C. (1999). On power-law relationships of the internet topology. In *SIGCOMM*, pages 251-262.

Gold, E. M. (1967). Language identification in the limit. *Information and Control*, 10:447-474.

Hagoort, P. (2003). How the brain solves the binding problem for language: a neurocomputational model of syntactic processing. *Neuroimage*, 20(Supplement 1):S18-S29.

Horning, J. J. (1969). *A study of grammatical inference*. PhD thesis, Stanford University, Stanford:CA, USA.

Hume, D. (1909). *An Enquiry Concerning Human Understanding*, volume XXXVII, Part 3 of *The Harvard Classics*. P.F. Collier & Son.

Huybrechts, R.M.A.C. (1984). The weak adequacy of context-free phrase structure grammar. In de Haan, G.J., Trommelen, M., and Zonneveld, W., editors, *Van periferie naar kern*, pages 81-99. Foris, Dordrecht, the Netherlands.

Ishiszaka, H. (1990). Polynomial time learnability of simple deterministic languages. *Machine Learning*, 5:151.

Kanazawa, M. (1995). *Learnable classes of categorial grammars*. PhD thesis, Stanford University, Stanford:CA, USA.

Lang, K. J., Pearlmutter, B. A., and Price, R. A. (1998). Results of the Abbadingo One DFA learning competition and a new evidence-driven state merging algorithm. In Honavar, V. and Slutzki, G., editors, Proceedings of the 4th International Conference on Grammar Inference, ICGI 1998, volume 1433 of LNCS/LNAI, pages 1-12. Springer-Verlag, Berlin Heidelberg, Germany.

Li, M. and Vitányi, P. M. B. (1991). Learning simple concepts under simple distributions. *SIAM Journal of Computing*, 20(5):911-935.

Osherson, D., de Jongh, D., Martin, E., and Weinstein, S. (1997). *Handbook of Logic and Language*, chapter Formal Learning Theory, pages 737-775. Elsevier Science B.V.

Pinker, S. (1999). *Words and Rules: The Ingredients of Language.* Weidenfeld and Nicolson, London, UK.

Pitt, L. and Warmuth, M. (1988). Reductions among prediction problems: On the difficulty of predicting automata. In *3rd Conference on Structure in Complexity Theory,* pages 60-69.

Sakakibara, Y. (1992). Efficient learning of context-free grammars from positive structural examples. *Information and Computation,* 97:23-60.

Seginer, Y. (2003). Learning context free grammars in the limit aided by the sample distribution. In de la Higuera, C., Adriaans, P., van Zaanen, M., and Oncina, J., editors, *Proceedings of the Workshop and Tutorial on Learning Context-Free Grammars held at the 14th European Conference on Machine Learning (ECML) and the 7th European Conference on Principles and Practice of Knowledge Discovery in Databases (PKDD); Dubrovnik, Croatia,* pages 77-88.

Shieber, S.M. (1985). Evidence against the context-freeness of natural language. Linguistics and Philosophy, 8(3):333-343.

Sokolov, J. and Snow, C. (1994). The changing role of negative evidence in theories of language development. In Gallaway, C. and Richards, B., editors, *Input and Interaction in Language Acquisition,* pages 38-55. Cambridge University Press, Cambridge, UK.

Solomonoff, R. J. (1997). The discovery of algorithmic probability. *Journal of Computer and System Sciences,* 55(1):73-88.

Valiant, L. G. (1984). A theory of the learnable. *Communications of the Association for Computing Machinery,* 27(11):1134-1142.

van Kampen, J. (1997). *First Steps in Wh-movement,* PhD thesis, Utrecht University, Utrecht, the Netherlands.

van Zaanen, M. (2002). *Bootstrapping Structure into Language: Alignment-Based Learning.* PhD thesis, University of Leeds, Leeds, UK.

van Zaanen, M. and Adriaans, P. (2001). Alignment-Based Learning versus EMILE: A comparison. In *Proceedings of the Belgian-Dutch Conference on Artificial Intelligence (BNAIC); Amsterdam, the Netherlands,* pages 315-322.

van Zaanen, M., Roberts, A., and Atwell, E. (2004). A multilingual parallel parsed corpus as a gold standard for grammatical inference evaluation. In Kranias, L., Calzolari, N., Thurmair, G., Wilks, Y., Hovy, E., Magnusdottir, G., Samiotou, A., and Choukri, K., editors, *Proceedings of the Workshop: The Amazing Utility of Parallel and Comparable Corpora; Lisbon, Portugal,* pages 58-61.

Vervoort, M. (2000). *Games, walks and Grammars.* PhD thesis, University of Amsterdam.

Vosse, T. and Kempen, G. (2000). Syntactic structure assembly in human parsing: a computational model on competitive inhibition and lexicalist grammar. *Cognition,* 75:105-143.

Wolff, J. G. (2003). Information Compression by Multiple Alignment, Unification and Search as a Unifying Principle in Computing and Cognition. *Journal of Artificial Intelligence Research,* 19:193-230.

Yokomori, T. (1988). Learning simple languages in polynomial time. Technical report, SIGFAI, Japanese Society for AI.

8 On Kernel Target Alignment

Nello Cristianini[1,] Jaz Kandola[2], Andre Elisseeff[3], and John Shawe-Taylor[4]

1. Department of Statistics, University of California at Davis,
   nello@support-vector.net
2. Merrill Lynch, Quantitative Analytics Division, 2 King Edward
   Street, London EC1A 1HQ, Jasvinder_Kandola@ml.com
3. IBM Research Lab, Saumerstrasse 4, CH-8803 Rueschlikon,
   Switzerland, AEL@zurich.ibm.com
4. School of Electronics and Computer Science, University of
   Southampton, UK, jst@ecs.soton.ac.uk

Abstract

Kernel based methods are increasingly being used for data modelling because of their conceptual simplicity and outstanding performance on many tasks. However, in practice the kernel function is often chosen using trial-and-error heuristics. In this paper we address the problem of measuring the degree of agreement between a kernel and a learning task. We propose a quantity to capture this notion, which we call alignment. We study its theoretical properties, and derive a series of simple algorithms for adapting a kernel to the targets. This produces a series of novel methods for both transductive and inductive inference, kernel combination and kernel selection for both classification and regression problems that are computationally feasible for large problems. The algorithms are tested on publicly available datasets and are shown to exhibit good performance.

8.1 Introduction

The development of methods that learn from data has been motivated by the need to discover implicit and non-trivial relationships existing in datasets. Kernel machines [9,19,13] work by first embedding the data into a feature space, and then by using linear algorithms to detect patterns in the images of the data. It is of crucial importance in such algorithms that the correct choice of kernel be made, since different kernels will generate different structures in this embedding space. As such being able to assess the quality of an embedding is an important problem from both a

N. Cristianini et al.: *On Kernel Target Alignment*, StudFuzz **194**, 205–256 (2006)
www.springerlink.com

theoretical and practical point of view, and one that has so far been treated only marginally.

In this paper we attempt to address this issue by proposing a measure of similarity between different embeddings of a set of data points. This quantity is called the *alignment* and can be used not only to assess the relationship between the embeddings generated by two different kernels, but also to assess the similarity between the embeddings of a labelled dataset induced by a kernel and that induced by the labels themselves.

The main theoretical contributions of this paper are in the definition of the problem of kernel-target similarity and in the proposal of the 'alignment' as a possible way to address it. We show that this quantity has several convenient properties: it can be efficiently computed before any training of the kernel machine takes place, and based only on training data information. A proof is included that shows alignment to be sharply concentrated around its expected value and hence its empirical value to be stable with respect to different splits of the data. This allows us to estimate it from finite samples, and to establish a connection between high alignment and good generalization performance. We also give a novel way to characterize a kernel function in terms of the alignment, and discuss some geometric interpretations of this quantity.

The main practical contribution is in giving a series of novel transductive and inductive learning algorithms in a number of settings. These include classification, regression, clustering and for the task of kernel combination. Particularly promising is the transductive approach, that provides a first step in the direction of 'nonparametric' kernel selection, in the sense that instead of choosing a parameterized kernel family and tuning the parameters so as to maximize the alignment (or other measures), we show how to directly adapt the entries of the kernel matrix to obtain a better one.

This paper is structured as follows. In Section 8.2 we review some basic concepts about kernel methods and provide a novel alternative characterization of kernel matrices. In Section 8.3 we propose the alignment as a measure of kernel similarity, and we derive some of its theoretical properties including its sharp concentration. In Section 8.5 and 8.6 we present a series of algorithms for kernel combination in both transductive and inductive settings for classification and regression problems.

Notations Vectors are generally written in lowercase letters: y stands for the target and is considered either as a finite dimensional vector or as a function from the input space $\mathcal{X}$ into $\mathbb{R}$. In the latter case, we write $y(x)$.

We denote by yy' the outer product between vector y and its transpose; and denote by $y \otimes y$ the function $y(x)y(z)$. When it is not possible to expressed the target as a function from the input space, we consider the conditional expectation $y(x) = \mathbb{E}[y \mid x]$. If y refers to a finite dimensional vector, it corresponds to a sampling of m points according to a fixed distribution P^m over the sample space $\mathcal{X} \times \mathbb{R}$. For classification, we will therefore consider $y \in \{-1, +1\}^m$ although $y(x)$ might be any number between $[-1, 1]$. The ith coordinate y_i is the target value for an input $x_i \in \mathcal{X}$ and the pairs (x_i, y_i), $i = 1, .., m$ form the elements of the training set that we denote S. The distribution P^m is assumed here to be the product of m identical distributions P over $\mathcal{X} \times \mathbb{R}$. For ease of notation, we will indistinctly use both P^m and P to denote P^m. When unclear from the context, we will sometimes use a subscript with P to indicate which random variable the probability applies to, e.g. P_S for a probability over the sampling of the training set S.

As we will compare matrices, we introduce here the Frobenius inner product $\langle .,. \rangle_F$ between two square matrices M and N of the same size:

$$\langle M, N \rangle_F = \sum_{ij} m_{ij} n_{ij} = Tr(MN)$$

When clear from the context, we will simply use $\langle M, N \rangle$ instead of $\langle M, N \rangle_F$.

For two-ary functions k and g over $\mathcal{X}$ we consider as well the following dot product

$$\langle k, g \rangle_P = \int_{\mathcal{X} \times \mathcal{X}} k(x, z)g(x, z)dP_\mathcal{X}(x)dP_\mathcal{X}(z)$$

where $P_\mathcal{X}$ the marginal distribution of P over $\mathcal{X}$. The corresponding norm will be denoted by $\| - \|_P$. Note that matrices are written in uppercase letters although we use lowercase letters for functions. Every ambiguity should be removed from the context. For instance in the equations $\langle vv', uu' \rangle_F = \langle v, u \rangle^2$ that we will use quite often, all lowercase letters refer to vectors. Finally, we assume that y and kernel values are all in the range $[-1, 1]$.

8.2 Similarity and Alignment

8.2.1 Kernels

Kernel based learning methods revolve around the notion of a "kernel matrix" K that can informally be regarded as a pairwise similarity matrix between points in a dataset. For a set of inputs points $\{x_1, x_2, ..., x_m\}$, this similarity depends solely on a function k called a kernel:

$$K = \left(k(x_i, x_j) \right)_{i,j=1}^{m} \tag{8.1}$$

Provided that this kernel matrix is computed, a bunch of very diverse algorithms can be applied: one class, two-class, multi-class classification, regression, etc. The choice of the algorithm depends directly on the problem at hand and hyperparameters are usually tuned with classical statistical techniques. What remains to the practitioner is the choice of the kernel.

Not every function can be interpreted as a kernel. We have the following condition:

Definition 1 [18] *A function* $k(x,z)$ *is a valid kernel iff for any finite set of   data* $\{x_1, x_2, ..., x_m\}$,   *for   any* $(a_1, .., a_m) \in \mathbb{R}^m$ *we   have*
$\sum_{i,j=1}^{m} a_i a_j k(x_i, x_j) \geq 0$.

For a particular application choosing a kernel corresponds to implicitly choosing a feature space. For every kernel k, there exists indeed a function ϕ from χ onto a potentially infinite dimensional vector space $\mathcal{F}$ endowed in with an inner product $\langle ., . \rangle_{\mathcal{F}}$ so that:

$$k(x,z) = \langle \phi(x), \phi(z) \rangle_{\mathcal{F}} \tag{8.2}$$

Any calculations can therefore be performed in $\mathcal{F}$ as long as it can be expressed in terms of dot products and so in terms of kernel functions. This is what kernel methods are all about: cast any algorithm into dot product calculations so that it can be applied to arbitrary vector spaces by replacing the dot product into a kernel function. Because kernels are dot products the kernel matrix K is also a Gram matrix in $\mathcal{F}$ and it is symmetric and definite positive.

It is interesting to consider the vector space formed by all the symmetric matrices of a given size Q_n, and in such a space the subset SPD_n of symmetric positive definite (SPD) matrices. We can provide the space Q_n with the Frobenius inner product defined in the previous section to induce a metric, and other structures. The subset of symmetric positive definite matrices SPD_n can then be shown to form a cone [14]:

Proposition 2 *Let K be a symmetric matrix. Then K is positive semi-definite if and only if $\langle K, G \rangle_F \geq 0$, for all positive semi-definite matrices G.*

Proof. Let $K = \sum_{i=1}^{m} \lambda_i v_i v'_i$ be the eigenvalue decomposition of K and $G = \sum_{j=1}^{m} \mu_j u_j u'_j$ that of a general G. We have

$$\langle K, G \rangle_F = \sum_{ij} \lambda_i \mu_j \langle v_i v'_i, u_j u'_j \rangle_F$$

$$= \sum_{ij} \lambda_i \mu_j (v'_i u_j)^2 \geq 0,$$

if $\lambda_i, \mu_j \geq 0$. Conversely, if $\lambda_i < 0$ for some i, choose $u_i = v_i$, $i = 1, ..., m$, and $\mu_j = 0$, for $j \neq i$, $\mu_i = 1$. It follows that $\langle K, G \rangle_F < 0$.

Notice that this proposition gives an alternative characterization of kernel functions.

8.2.2 Learning and Similarities

Choosing a kernel reduces to choosing a representation and the latter is not obvious: it might be easy to encode numbers or colors, but it's much more difficult to encode a concept like a string or a person. The reason might be that we would like the encoding to capture some of the relation that we have to learn: two inputs from the same group should be close together. If this is not the case, generalization seems hopeless.

In a way, learning might be interpreted as learning similarities: two inputs are similar if they are from the same class. Every classification model can as well be interpreted as similarity measure: two points are similar if the model gives the same output to both. Finding a good learning algorithm amounts then to finding a similarity measure that fits well the similarity encoded in the target function y. From that perspective, it is interesting to see what would be a cost function that would penalize similarities that are far away from the concept to learn.

If the targets y are encoded as ± 1, a natural similarity measure between x and z is the similarity $y(x)y(z)$ so that is it equal to 1 if they are from the same class and -1 otherwise. For a model f from χ to $\{-1,+1\}$, a similarity would as well be $f(x)f(z)$. A somewhat direct definition of a fit between similarity measures is then:

$$\int_{\chi \times \chi} (f(x)f(z) - y(x)y(z))^2 \, dP_\chi(x)dP_\chi(z)$$

When the input space χ is finite, functions f and y are vectors and the above measure is the Frobenius distance between ff' and yy'.

The similarity we consider here is directly built from a model but the choice of a "good" representation could then be addressed independently of the algorithm that will be built from it. This might appear as a default since it will not take into account the specificity of the learning algorithm and has then no reason to give good results. On the other hand, it might avoid overfitting since the goal is not then to improve the performance of one particular learning algorithm. This is very similar to the feature selection discussion about filter and wrapper approaches: no clear advantages for one or the other method has been stated and using correlation between features and target is still used as any other wrapper method.

As we have seen before, a representation is implicitly built from a dot product that we call a kernel $k(x,z)$. The latter can be interpreted as similarities so that we can define the distance between a kernel/representation k and the target y as:

$$\int_{\chi \times \chi} (k(x,z) - y(x)y(z))^2 \, dP(x)dP(z)$$

As noted previously, if χ is finite, then k is a matrix and this distance is the same as the Frobenius norm between k and the matrix $y(x)y(z)$. The problem with this distance is that it is sensitive to scaling: if the target is encoded as $\{-2,+2\}$ the distance will change, which is not a desirable property. To avoid this, we normalize the similarity measure by their norm and compute the new "normalized" distance:

$$\int_{\chi \times \chi} \left(\frac{k(x,z)}{\|k\|_P} - \frac{y(x)y(z)}{\|y \otimes y\|_P} \right)^2 dP_\chi(x) dP_\chi(z) = 2 - \frac{\langle k, y \otimes y \rangle_P}{\|k\|_P \|y \otimes y\|_P} \tag{8.3}$$

Note that one could as well center the similarities such that if the target is encoded as $\{-1/2, 3/2\}$ the distance does not change. This however induces more notations and a heavier theoretical analysis. Since the alignment will be used mainly to tune kernels, this defect might be corrected by always adding a bias to kernels and by maximizing the alignment with respect to this bias as well as with respect to the kernel. In this case, it is easily seen that whether y has a shift or not will not change the final optimized alignment.

The left hand side of (8.3) is computed via a quantity that is very close to a correlation measure in the product space $\chi \times \chi$ between two random variables $k(x,z)$ and $y(x)y(z)$. Correlation would require centered kernels which is the not the case here but as we said is very in the spirit of what we would like to measure.

8.2.3 Definition of the Alignment

We call the correlation between two similarities derived from kernels the *alignment* and define:

Definition 3 (Alignment) *Let k_1 and k_2 be two kernels, we define the alignment between k_1 and k_2 as:*

$$A(k_1, k_2) = \frac{\langle k_1, k_2 \rangle_P}{\|k_1\|_P \|k_2\|_P} \tag{8.4}$$

In particular, when the kernel k_2 is built from the target function $y(x)$, it becomes:

$$A(k_1, y) = \frac{\langle k_1, y \otimes y \rangle_P}{\|k_1\|_P \|y \otimes y\|_P} \tag{8.5}$$

When χ is finite, this alignment is the cosine between the matrices K_1 and K_2 derived from k_1 and k_2. This quantity is therefore maximized when $K_1 \propto K_2$. It is worth mentioning here that the alignment is built from the canonical dot product in the Euclidean or Hilbertian product space $\mathcal{X} \times \mathcal{X}^{1}$. In that space k_1 and k_2 or $y \otimes y$ are vectors and the Frobenius norm that we mention previously is the ℓ_2 norm. One could define another fit or notion of alignment for similarities based on another norm on this space, add prior knowledge and focus only on subsets of pairs (x, z). We did not consider such approach here and define the alignment with the idea to introduce as few bias as possible: no pair of points is favored/weighted except by its probability to occur.

Computing the alignment to a target requires having total knowledge of the target function, while in a real problem one only has access to this information for the points in the training set. So only the *empirical* alignment can be directly measured, which correspond to the case where $\mathcal{X}$ is finite and is limited to the inputs of the training set $x_1, .., x_m$. The empirical alignment between the kernel matrix K and the target matrix $Y = (y_i y_j)_{i, j=1, .., m}$ can be defined as:

Definition 4 (Empirical Alignment) *Let K_1 and K_2 be two kernel matrices derived from two kernels k_1, k_2 from a training set $S = \{(x_1, y_1), .., (x_m, y_m)\}$. We define the empirical alignment between these two kernels on S as:*

$$\hat{A}(k_1, k_2, S) = \frac{\langle K_1, K_2 \rangle_F}{\|K_1\|_F \|K_2\|_F} \tag{8.6}$$

When k_2 is derived from the target $y(x)$, it becomes:

$$\hat{A}(k, y, S) = \frac{\langle K, yy' \rangle_F}{\|K\|_F \|yy'\|_F} \tag{8.7}$$

where $= (_1, .., _m)^T$.

Note that while introducing the alignment we only talked about classification but the same reasoning could be used for regression: similarities in the input space should reflect in a way similarities in the output space. The latter might be represented by the product $y \otimes y$ but might as well be defined differently. One could indeed think about $exp(-(y(x) - y(z))^2)$. Using non-trivial similarities in the output space is not addressed in this paper. The main issue that we will discuss in S

---

[1]We assume that $\mathcal{X}$ is a Hilbert space. Practically, it is almost always the case.

Section 8.4 is how to use the alignment to perform kernel selection and to deduce new learning algorithms.

8.3 Properties of the Alignment

The range of the alignment is $[0,1]$. The larger its value the closer is the kernel from the target. We present in this section a short discussion about kernels, prior knowledge and their relation with the alignment. In a second part, we explore the statistical properties of the empirical alignment.

8.3.1 Building kernels with large alignment

Ideally the kernel used for a specific learning task should induce a structure in the feature space as close as possible to the relations to be learned from data, so to facilitate the work of the learning algorithm. In other words, the target function to be learned should be somewhat prominently (although possibly implicitly) represented in the features defined by the kernel function. Of course, this information is unknown a priori, but often domain knowledge is present that can at least guess a subset of functions that should contain the target, and use that information to design a kernel. If the unknown target turns out to be part of the expected class, good performance should be expected.

Let us assume that the target is chosen from a set $\mathcal{T}$ of targets (from X to $\{-1,+1\}$) that we know beforehand. This prior knowledge should guide us to define a kernel whose alignment with the target is large. We start with the trivial case where cardinal of $\mathcal{T}$ is equal to 1.

Example 5 ($|\mathcal{T}|=1$) *In that case the ideal feature vector is given by the desired classification, that is*

$$\phi : x \mapsto y(x). \tag{8.8}$$

The corresponding kernel function is given by

$$k(x,z) = y(x)y(z), \tag{8.9}$$

and we can now learn the target from one example (x_1, y_1) as

$$f(x) = y_1 k(x_1, x) = y_1 y(x_1) y(x) = y(x). \tag{8.10}$$

Note that we can also learn the negation of this function. The alignment is equal to 1 which is the largest possible value for an alignment.

In this example, any non stupid learning algorithm would learn the target perfectly. Now consider a slightly more complex case:

Example 6 ($|T| = 2$) *For two functions y^1 and y^2 the obvious feature space gives rise to the kernel*

$$k(x,z) = y^1(x)y^1(z) + y^2(x)y^2(z), \qquad (8.11)$$

in which as many as 8 distinct functions can be realized (in general the four points $(-1,-1)$, $(-1,1)$, $(1,-1)$ and $(1,1)$ will be images of some input points in the feature space $\{(y_1(x), y_2(x)) : x \in X\}$ and there are eight dichotomies of these points that can be realized). The alignment is then the same between this kernel and any of the two target y_1 or y_2 so that they are considered equally. If we had considered a convex combination $k(x,z) = ty^1(x)y^1(z) + (1-t)y^2(x)y^2(z)$ for $t \in [0,1]$, then alignments would have been different and would have favored the target with the largest value over t or $1-t$.

More generally we expect that a kernel defined from different possible targets will bias the alignment toward the target to which it gives the maximum weight.

Example 7 ($|T| < \infty$) *Assume that T is formed by a finite number of target functions y^l that are orthogonal with respect to the inner product,*

$$\langle y^1, y^2 \rangle_P = \int_X y^1(x)y^2(x)dP_X(x), \qquad (8.12)$$

We can consider the kernel:

$$k = \sum_l \mu_l(y^l \otimes y^l), \qquad (8.13)$$

and rewrite it as

$$k = \mu_t(y^t \otimes y^t) + \sum_{l \neq t} \mu_l(y^l \otimes y^l) \qquad (8.14)$$

We can now evaluate the alignment as follows:

$$A = \frac{\langle k, y^t \otimes y^t \rangle_P}{\sqrt{\langle k,k \rangle_P \langle y^t \otimes y^t, y^t \otimes y^t \rangle_P}} = \frac{\sum_l \mu_l \langle y^l \otimes y^l, y^t \otimes y^t \rangle_P}{\sqrt{\sum_{i,j} \mu_i \mu_j \langle y^i \otimes y^i, y^j \otimes y^j \rangle_P}} \qquad (8.15)$$

But observe that

$$\langle y^i \otimes y^i, y^j \otimes y^j \rangle_P = \langle y^i, y^j \rangle_P^2 = \delta_{ij} \qquad (8.16)$$

so that the alignment will be simply

$$A = \frac{\mu_t}{\sqrt{\sum_l \mu_l^2}} \qquad (8.17)$$

When χ is finite, k is a matrix K, namely a Gram matrix, and can be written as: $K = \sum \lambda_i v_i v_i'$ where the v_i's are its eigenvectors. The kernel can then be interpreted as a combination of similarities derived from targets given by the eigenvectors. The latter are weighted by the eigenvalues. Therefore, the more peaked the spectrum the more aligned (specific) the kernel can be.

One could easily see the analogy with Gaussian kernel: if we consider kernels defined as,

$$k(x,z) = \int_C c(x)c(z)d\mu(c), \tag{8.18}$$

where μ is a prior probability distribution over the function class C, then the alignment with a target y will give the largest value when $\mu(c)$ is peaked around y. Note that [2] have also studied a related concept in assessing the realisability of function classes using kernels.

8.3.2 Statistical Properties of the Alignment

We prove here that empirical alignment can be used to tune the true alignment. The reasoning is based on a concentration inequality of the empirical alignment around its expectation. This inequality will play the same role as Hoeffding's bound for deviation between empirical mean and true expectation. We will prove that the empirical alignment defined as:

$$\hat{A}(k_1, k_2, S) = \frac{\langle K_1, K_2 \rangle_F}{\|K_1\|_F \|K_2\|_F}$$

(where $K_l = (k_l(x_i, x_j))_{i,j=1,..,m}$ for $l = 1, 2$) is very close to its expectation:

$$\mathbb{E}\left[\frac{\langle K_1, K_2 \rangle_F}{\|K_1\|_F \|K_2\|_F} \right]$$

Note that the expectation is over the training set $(x_1, .., x_m)$. We will then prove that this expectation is close to the *true* alignment. For this purpose, we need to introduce the following theorem:

Theorem 8 ([15]) *Let $X_1,...,X_n$ be independent random variables taking values in a set A, and assume that $f : A^n \to \mathbb{R}$ satisfies for $1 \le i \le n$*

$$\sup_{x_1,...,x_n,\hat{x}_i} | f(x_1,...,x_n) - f(x_1,...,x_{i-1},\hat{x}_i,x_{i+1},...,x_n) | \le c_i,$$

then for all $\epsilon > 0$,

$$P\{| f(X_1,...,X_n) - \mathbb{E}f(X_1,...,X_n) | \ge \epsilon\} \le 2\exp\left(\frac{-2\epsilon^2}{\sum_{i=1}^n c_i^2}\right)$$

Let us denote by S^r the training set $S = \{(x_1,y_1),..,(x_m,y_m)\}$ where one point (x_r,y_r) has been replaced by (x'_r,y'_r). Then using Proposition 17 (proved in the appendix), we have:

$$\left|\hat{A}(k,h,S) - \hat{A}(k,h,S^r)\right| \le \frac{12}{ma^2} \tag{8.19}$$

which can be plugged in McDiarmid's theorem to get a concentration result for the empirical alignment:

Theorem 9 *Let k and h be two kernels and $S = \{x_i\}_{i=1,..,m}$ a set of input points. Consider $K = (k(x_i,x_j))_{i,j=1,..,m}$ and $H = (k(x_i,x_j))_{i,j=1,..,m}$ the values of k and h on the finite set S. Assume that $\|H\|$ and $\|K\|$ are greater than a for any S, then:*

$$P\left(|\hat{A}(k,h,S) - \mathbb{E}[\hat{A}(k,h,S)| \ge \epsilon\right) \le 2\exp\left(-\frac{ma^4\epsilon^2}{c_1}\right) \tag{8.20}$$

where c_1 is a constant.

Now the question is whether the expectation of the empirical alignment is close to the true alignment. The answer is positive and relies on the concentration of $\|K\|_F^2 = \frac{1}{m^2}\sum_{i,j} k^2(x_i,x_j)$ around its expectation. With probability $1 - \delta$, we have:

$$\left\| \|K\|_F^2 - \mathbb{E}\left[\|K\|_F^2\right]\right\| \le \sqrt{\frac{c_2 ln(2/\delta)}{m}} \tag{8.21}$$

where c_2 is a constant. This inequality comes from theorem 8 applied to K. Note that the expectation is slightly different from $\mathbb{E}_{x,z}\left[k^2(x,z)\right]$

because of the terms $k^2(x_i, x_i)$ in the expression of K. However, these terms contribute only for $1/m$ of the total value of K and we will neglect them. We have then with probability $1 - \delta$:

$$\left| \hat{A}(h,k,S) - \frac{(1/m^2)\sum_{i,j} k(x_i,x_j)h(x_i,x_j)}{\sqrt{\mathbb{E}\left[k^2(x,z)\right]}\sqrt{\mathbb{E}\left[h^2(x,z)\right]}} \right| \leq \frac{1}{a^2}\sqrt{c_2 \frac{ln(2/\delta)}{m}}$$

where we assume here that $1/m^2 \sum_{i,j} h^2(x_i,x_j)$ and $1/m^2 \sum_{i,j} k^2(x_i,x_j)$ are always greater than a^2. Taking the expectation of both side of the previous equation and setting $\delta = 1/m$, we derive:

$$\left| \mathbb{E}\left[\hat{A}(k,h,S)\right] - A(k,h) \right| \leq \frac{1}{m} + \frac{1}{a^2}\sqrt{c_2 \frac{ln(2m)}{m}} \qquad (8.22)$$

We have now a proof that under certain condition the empirical alignment is close to the true alignment. The problem is that this closeness holds only for fixed kernel k and h. If the alignment is to be optimized with respect to k for instance, the result we just showed does not tell how the alignments will behave. To have such information, we need to derive bounds on the alignment uniformly for a family of kernels. The latter can be defined as a set K of 2-ary functions from $X \times X$ to $\mathbb{R}$. We are thus interested in bounding:

$$\sup_{k \in K} \left| \hat{A}(k,h) - \mathbb{E}\left[\hat{A}(k,h)\right] \right|$$

where h is any fixed kernel or the kernel induced by the target function. Note that $\mathbb{E}\left[\hat{A}(k,y,S)\right]$ is slightly different from the true alignment $A(k,y)$ but as we shown previously, it is not very far away. When m tends to infinity both quantities will be equal if $\|k\|$ and $\|y\|$ are greater than a constant that we denoted a. We have the following theorem:

Theorem 10 *Let K a set of kernels, $a \in \mathbb{R}$ and h be a fixed kernel. Assume that for all $k \in K$, for any $S = \{x_1,..,x_m\}$ in X^m, $\|K\|_S = \frac{1}{m^2}\sum_{i,j=1}^m k^2(x_i,x_j) \geq a$ and that $\|H\|_S \geq \|a\|$, then we have:*

$$P_S \left[\sup_{k \in K} \left| \hat{A}(h,k,S) - \mathbb{E}[\hat{A}(h,k,S)] \right| \geq \epsilon \right] \leq$$

$$2\mathbb{E}\left[\mathcal{N}\left(a\epsilon/(18\pi), \mathcal{K}, d_{S,S'} \right) \right] \exp\left(-\frac{ma^4\epsilon^2}{c_2} \right) \tag{8.23}$$

where S' is a second set $\{x_{m+1},..,x_{2m}\}$ sampled identically to S and $d_{S,S'}(h,k)$ is defined as:

$$d_{S,S'}(h,k) = \max_{\sigma} \sqrt{ \frac{1}{m^2} \sum_{i,j=1}^{m} (h(x_{\sigma(i)},x_{\sigma(j)}) - k(x_{\sigma(i)},x_{\sigma(j)}))^2 + }$$

$$\sqrt{ \frac{1}{m^2} \sum_{i,j=m}^{2m} (h(x_{\sigma(i)},x_{\sigma(j)}) - k(x_{\sigma(i)},x_{\sigma(j)}))^2 } \tag{8.24}$$

where σ is any permutation between $\{1,..,m\}$ and $\{m+1,..,2m\}$ swapping the x_i with the x_{m+i}'s: $\sigma(i) \in \{i, m+i\}$. c_2 is a constant.

This theorem gives a bound similar to what is derived for the Empirical Risk Minimization principle. The difference is here that Hoeffding's bound has been replaced by equation (8.20). The requirement for the latter to be applied is that all kernels are uniformly lower bounded by a. This is a strict condition that might be forced by adding a constant to the kernels. We see now why centering might not be good from a theoretical point of view: it might provide a better measure of similarity but the latter could not then be controlled as conveniently. In the next section we will see how maximizing alignment can give simple algorithms for selecting kernel parameters and for combining kernels. It is worth mentioning then that the norm of the resulting kernel must be lower bounded during this maximization so that the empirical alignment can be trusted. Beside this lower bound, one needs as well to upper bound the covering number of $\mathcal{K}$ with respect to some metric. This can be done as in the following example.

Example 11 (Covering number bound for convex combinations of kernels)

Let $S = \{x_1,..,x_m\}$ and $S' = \{x_{m+1},..,x_{2m}\}$. Consider convex combinations of fixed kernels $k_1,..,k_T$:

$$K = \{ \sum_{t=1}^{T} \lambda_t k_t \text{ s.t. } \sum_{t=1}^{T} \lambda_t = 1 \text{ and } \lambda_t \geq 0 \}$$

and embed $\mathcal{K}$ in the product space $\mathbb{R}^{4m^2}$ with the averaged $\ell_{a,2}(S,S')$ inner product:

$$\langle k,h \rangle_{\ell_{a,2}(S,S')} = \frac{1}{m^2} \sum_{i,j=1}^{2m} k(x_i,x_j)h(x_i,x_j)$$

Then: $d_{S,S'}(k,h) \leq \| k-h \|_{\ell_{a,2}(S,S')}$ and

$$\mathcal{N}\left(a\epsilon/(18\pi),\mathcal{K},d_{S,S'}\right) \leq \mathcal{N}\left(a\epsilon/(18\pi),\mathcal{K},\ell_{a,2}(S,S')\right)$$

To bound the right hand side, we use the evaluation function:

$$E_{S,S'} : (\mathbb{R}^T,\ell_1) \longrightarrow \left(\mathbb{R}^{4m^2},\ell_{a,2}(S,S')\right)$$

$$(\lambda_1,..,\lambda_T) \longmapsto \left(\sum_{t=1}^{T} k_t(x_i,x_j)\right)_{i,j=1,..,2m}$$

and note that the range of $E_{S,S'}$ contains $\mathcal{K}$ so that the covering number of $\mathcal{K}$ in $\left(\mathbb{R}^{4m^2},\ell_{a,2}(S,S')\right)$ in upper bounded by the covering number of the range of $E_{S,S'}$. But the latter can be bounded because $E_{S,S'}$ is a linear operator. Using results from [6], we get:

$$\mathcal{N}\left(a\epsilon/(18\pi),\mathcal{K},d_{S,S'}\right) \leq \mathcal{N}\left(a\epsilon/(18\pi),\mathcal{K},\ell_{a,2}(S,S')\right) \leq \left(\frac{36\pi}{a\epsilon}\right)^T$$

The bounds derived from theorem 10 are not to be used directly in practice. They just give hints about how one might optimize the alignment. We can in particular see that if the capacity of the set of kernels, measured as a covering number, is bounded then at least for large value of m, the empirical alignment can be used to maximize the true alignment. As discussed before, this conclusion holds when the norm of the kernels are greater than a constant.

8.3.3 Alignment and Generalization

The remaining question that has not yet been answered except by a discussion is whether it makes sense to maximize the true alignment. It might be relevant to model the similarities encoded in the target function y but it is not clear yet whether it helps generalization or not. We prove existence of a hypothesis that is not computable from the sample but its existence implies that good generalization can at least in principle be obtained with the given feature space. That is we can prove that if the alignment is high then there always exists a good classification hyperplane. Consider the function

$$f(x) = \frac{\int_X y(z)k(x,z)dP_\chi(z)}{\|k\|_P},$$

(8.25)

where $y(z)$ are the true labels. Note that $-1 \le f(x) \le 1$. Let $m(x) = y(x)f(x)$ be the margin of f on the point x. Now we have that

$$\mathbb{E}m(x) = \frac{\int_X y(x)f(x)dP\chi(x)}{\|k\|_P}$$

$$= \frac{\int_X y(x)\int_X y(z)k(z,x)dP_\chi(z)dP_\chi(x)}{\|k\|_P}$$

$$= \frac{\int_{X^2} y(x)y(z)k(x,z)dP_\chi(x)dP_\chi(z)}{\|k\|_P}$$

$$= A(y)$$

If $f(x)$ has generalization accuracy ϵ^+ and error ϵ^- then the largest value of the expected margin is when all of the correctly classified points have margin 1 and all of the misclassified points have margin 0. Hence,

$$A(y) \le 1 \times \epsilon^+ + 0 \times \epsilon^- = \epsilon^+,$$

(8.26)

so that $\epsilon^- \le 1 - A(y)$. The following theorem follows readily.

Theorem 12 *Let S be a set of input points of size m. Given any $\delta > 0$, with probability $1 - \delta$, the generalization accuracy of the function*

$$f(x) = \frac{\int_X y(z)k(x,z)dP(z)}{\|k\|_P},$$

(8.27)

is bounded from below by

$$\hat{A}(k, y, S) - \epsilon,$$

where

$$\epsilon = O\left(\frac{1}{a^2}\sqrt{\frac{\ln(m\delta)}{m}}\right)$$

(8.28)

Proof First observe that by Theorem 9 with probability greater than $1 - \delta$,

$$| A(k,y,S) - \mathbb{E}\left[\hat{A}(k,y,S)\right]| \leq \frac{1}{a}\sqrt{\frac{c_1 ln(2/\delta)}{m}} \qquad (8.29)$$

and according to equation (8.22) we have:

$$\left|\mathbb{E}\left[\hat{A}(k,y,S)\right] - A(k,y)\right| \leq \frac{1}{m} + \frac{1}{a^2}\sqrt{c_2 \frac{ln(2m)}{m}} \qquad (8.30)$$

So that we finally have:

$$\left|A(k,y,S) - A(k,y)\right| \leq \frac{1}{m} + \frac{1}{a^2}\sqrt{c_2\frac{ln(2m)}{m}} + \frac{1}{a}\sqrt{\frac{c_1 ln(2/\delta)}{m}} = O\left(\frac{1}{a^2}\sqrt{\frac{ln(m\delta)}{m}}\right)$$

Now by the above reasoning with probability greater than $1 - \delta$ the generalization accuracy of f can be lower bounded by

$$A(y) = A(k,y) \geq \hat{A}(S) - \epsilon_A, \text{ with } \epsilon = O\left(\frac{1}{a^2}\sqrt{\frac{ln(m\delta)}{m}}\right).$$

An empirical estimate of the function f would be the Parzen window estimate

$$F(x) = \frac{1}{m}\sum_{i=1}^{m} y_i k(x_i, x). \qquad (8.31)$$

By showing that this function is drawn from a class with bounded complexity (for example bounded fat-shattering dimension), it is possible to show that the function F converges uniformly to the function f, hence obtaining a bound on the generalization of F in terms of the empirically estimated alignment $\hat{A}(S)$. The concentration of F is considered in [10].

8.4 Algorithms for Alignment Optimization

The results obtained in the previous sections have several practical implications. In the next subsections we will examine some algorithms based on the concept of alignment between kernel and target. They can be used for kernel selection and combination; for unsupervised and transductive learning; and for kernel adaptation.

8.4.1 Adapting the Alignment

We now investigate how a kernel matrix can be adapted to improve the alignment with a target classification. We consider the general case where not all of the classifications are known. Assume that there are a subset L of indices for which target values are known. The method we propose is based on solving an optimization problem.

Optimizing the spectrum. This method can be used for transduction, by combining it with the techniques discussed elsewhere in this paper. Consider as base kernels the following $K_i = v_i v_i'$ where v_i are the eigenvectors of K, the kernel matrix for both labeled and unlabeled data. We define the parametrized class of kernels determined by this equation:

$$\hat{K} = \sum_i \alpha_i v_i v_i' \qquad (8.32)$$

and we consider the optimization problem of finding the optimal α, that is the parameters that maximize the alignment of the combined kernel with the available labels. Given $K = \sum_i \alpha_i v_i v_i'$, the alignment can be written as

$$A = \frac{\langle K, YY' \rangle}{m \sqrt{\sum_{ij} \alpha_i \alpha_j \langle v_i v_i', v_j v_j' \rangle}} \qquad (8.33)$$

From the orthonormality of the v_i and since $\langle vv', uu' \rangle = \langle v, u \rangle^2$ we can write:

$$A = \frac{\sum_i \alpha_i \langle v_i, y \rangle^2}{\sqrt{\langle yy', yy' \rangle} \sqrt{\sum_i \alpha_i^2}} \qquad (8.34)$$

Hence we have the following optimization problem: maximize

$$W(\alpha) = \sum_i \alpha_i \langle v_i, y \rangle^2 - \lambda(\sum_i \alpha_i^2 - 1) \qquad (8.35)$$

$$\frac{\partial W}{\partial \alpha_i} = \langle v_i, y \rangle^2 - \lambda 2\alpha_i = 0 \qquad (8.36)$$

and hence

$$\alpha_i \propto \langle v_i, y \rangle^2 \qquad (8.37)$$

This gives the overall alignment:

$$A = \sqrt{\frac{\sum_i \langle v_i, y \rangle^4}{\langle yy', yy' \rangle_F}} \qquad (8.38)$$

A transduction algorithm can be designed to take advantage of this, by optimizing alignment with the labeled part of the dataset, and in doing so it will adapt the Gram matrix also for the unlabeled part. The use of techniques like the spectral labeling method described in this paper can then provide a partitioning or a clustering of the data.

To demonstrate the performance of these algorithms we have used two binary classification datasets. The ionosphere dataset[2] which contains 34 inputs, a single binary output and 351 datapoints. The Wisconsin breast cancer dataset was obtained from the University of Wisconsin hospitals. It contains nine integer valued inputs, a single binary output (benign or malignant tumours) and 699 datapoints [3].

Three learning algorithms were implemented. A Parzen window estimator, a support vector classifier (SVM) and a v-SVM. A 10-fold procedure was used to find the optimal values for the capacity control parameters 'C' and 'nu' for a range of values of 'C' and 'nu'[3]. The SVM and v-SVM were trained ten times using a range of values of 'C' and 'nu'. The value of 'C' and 'nu' which gave the lowest mean error (and associated standard deviation) on a validation dataset was chosen as the optimal value. Having selected the optimal values of 'C' and 'v', the SVM and v-SVM were re-trained twenty times using twenty random data splits.

We applied the transduction algorithm designed to take advantage of our results by optimizing alignment with the labeled part of the dataset using the spectral method described in this paper, and so adapting the Gram matrix also to the unlabeled part. All of the results are averaged over twenty random data partitions with the standard deviation given in

---

[2]available from the UCI data repository
[3]The values considered for 'C' and 'nu' respectively were [0.01 10 100 500 1000 5000 10000 25000 50000] and [0.1 0.2 0.3 0.4 0.5 0.6 0.7 0.8 0.9]

brackets. Tables 1,2,3 and 4 show the mean generalisation error and associated standard deviation (in brackets) for the optimal value of 'C' or 'ν' for the datasets and data partitions considered.

The alignment algorithm was applied to the Breast cancer and Ionosphere datasets using either a Gaussian or linear kernel. Tables 5 and 6 show the alignments of the Gram matrices to the label matrix for different sizes of training set. The index indicates the percentage of training points. The K matrices are before adaptation, while the G matrices are after optimization of the alignment. The left two columns of Table 5 shows the alignment values for Breast Cancer data using a Gaussian kernel together with the performance of an SVM classifier and ν-SVM trained with the given gram matrix in the third column. The right two columns show the performance of the Parzen window classifier on the test set for Breast linear kernel (left column) and Ionosphere (right column). The theory did not make any prediction about SVM performance, but it is apparent from the table that for both 80% and 50% training sets there is a reduction in the generalisation error.

	Breast Cancer Dataset Linear Kernel	
	Test Error	Parameter Value
K_{80}	0.128 (0.03)	C = 10000
G_{80}	0.121 (0.02)	C = 25000
K_{80}	0.337 (0.03)	ν = 0.5
G_{80}	0.131 (0.02)	ν = 0.8
K_{50}	0.126 (0.02)	C = 10000
G_{50}	0.122 (0.01)	C = 25000
K_{50}	0.336 (0.03)	ν = 0.5
G_{50}	0.139 (0.02)	ν = 0.9
K_{20}	0.136 (0.01)	C = 5000
G_{20}	0.135 (0.02)	C = 5000
K_{20}	0.334 (0.03)	ν = 0.5
G_{20}	0.128 (0.02)	ν = 0.7

Table 1: Optimal values of the capacity control parameters for an SVM and ν-SVM. The mean (and standard deviation) of the estimated generalisation error over ten runs is quoted for the different data partitions into training and test sets.

The results clearly show that optimizing the alignment on the training set does indeed increase its value in all but one case by more than the sum of the standard deviations. Furthermore, as predicted by the concentration this improvement is maintained in the alignment measured on the test set with both linear and Gaussian kernels in all but one case (20% train with the linear kernel). The results for Ionosphere are less conclusive. Again as predicted by the theory the larger the alignment the better the performance that is obtained using the Parzen window estimator. The results of applying an SVM to the Breast Cancer data using a Gaussian kernel show a very slight improvement in the test error for both 80% and 50% training sets.

	Breast Cancer Dataset Gaussian Kernel	
	Test Error	Parameter Value
K_{80}	0.140 (0.03)	$C = 500$
G_{80}	0.133 (0.03)	$C = 1000$
K_{80}	0.220 (0.07)	$\nu = 0.6$
G_{80}	0.153 (0.03)	$\nu = 0.9$
K_{50}	0.146 (0.01)	$C = 5000$
G_{50}	0.143 (0.01)	$C = 25000$
K_{50}	0.204 (0.02)	$\nu = 0.7$
G_{50}	0.152 (0.03)	$\nu = 0.9$
K_{20}	0.147 (0.01)	$C = 100$
G_{20}	0.146 (0.05)	$C = 100$
K_{20}	0.251 (0.05)	$\nu = 0.7$
G_{20}	0.178 (0.03)	$\nu = 0.9$

Table 2: Optimal values of the capacity control parameters for an SVM and ν-SVM. The mean (and standard deviation) of the estimated generalisation error over ten runs is quoted for the different data partitions into training and test sets.

	Ionosphere Dataset Linear Kernel	
	Test Error	Parameter Value
K_{80}	0.164 (0.04)	C = 100
G_{80}	0.153 (0.03)	C = 5000
K_{80}	0.234 (0.04)	v = 0.5
G_{80}	0.119 (0.08)	v = 0.8
K_{50}	0.164 (0.02)	C = 100
G_{50}	0.156 (0.05)	C = 1000
K_{50}	0.292 (0.02)	v = 0.5
G_{50}	0.131 (0.02)	v = 0.8
K_{20}	0.202 (0.03)	C = 100
G_{20}	0.196 (0.03)	C = 500
K_{20}	0.346 (0.04)	v = 0.6
G_{20}	0.166 (0.05)	v = 0.8

Table 3: Optimal values of the capacity control parameters for an SVM and v-SVM. The mean (and standard deviation) of the estimated generalisation error over ten runs is quoted for the different data partitions into training and test sets.

	Breast Cancer Dataset		Ionosphere Dataset	
	Train Align	Test Align	Train Align	Test Align
K_{80}	0.076 (0.007)	0.092 (0.029)	0.208 (0.017)	0.232 (0.066)
G_{80}	0.228 (0.012)	0.219 (0.041)	0.233 (0.015)	0.244 (0.062)
K_{50}	0.075 (0.016)	0.084 (0.017)	0.211 (0.033)	0.215 (0.034)
G_{50}	0.242 (0.023)	0.181 (0.043)	0.246 (0.026)	0.214 (0.026)
K_{20}	0.072 (0.022)	0.081 (0.006)	0.227 (0.057)	0.210 (0.014)
G_{20}	0.273 (0.037)	0.034 (0.046)	0.277 (0.051)	0.183 (0.023)

Table 4: Mean and associated standard deviation alignment values using a linear kernel on the Breast Cancer and Ionosphere datasets over twenty runs.

| Alignment Breast Cancer Dataset | | SVM Error | ν SVM Error |
Train Align	Test Align	Test Error	Test Error
K_{80} 0.263 (0.011)	0.242 (0.041)	0.160 (0.023)	0.196 (0.059)
G_{80} 0.448 (0.012)	0.433 (0.047)	0.158 (0.027)	0.153 (0.035)
K_{50} 0.196 (0.015)	0.198 (0.016)	0.146 (0.008)	0.204 (0.061)
G_{50} 0.459 (0.017)	0.463 (0.017)	0.146 (0.007)	0.159 (0.022)
K_{20} 0.251 (0.048)	0.260 (0.012)	0.147 (0.010)	0.245 (0.097)
G_{20} 0.448 (0.055)	0.441 (0.014)	0.146 (0.006)	0.162 (0.029)

Table 5: Breast Cancer dataset: alignment values, SVM error and ν-SVM error for a Gaussian kernel (sigma = 6) over twenty runs

| Alignment Ionosphere Dataset | | SVM Error | ν-SVM Error |
Train Align	Test Align	Test Align	Test Align
K_{80} 0.208 (0.017)	0.232 (0.066)	0.164 (0.046)	0.141 (0.035)
G_{80} 0.233 (0.015)	0.244 (0.062)	0.158 (0.042)	0.334 (0.074)
K_{50} 0.211 (0.033)	0.215 (0.034)	0.184 (0.021)	0.141 (0.016)
G_{50} 0.246 (0.026)	0.214 (0.026)	0.167 (0.025)	0.348 (0.039)
K_{20} 0.227 (0.057)	0.210 (0.014)	0.204 (0.032)	0.165 (0.023)
G_{20} 0.277 (0.051)	0.183 (0.023)	0.186 (0.025)	0.355 (0.015)

Table 6: Ionosphere dataset: alignment values, SVM error and ν-SVM for a linear kernel over twenty runs.

| | Breast Cancer Dataset | Ionosphere Dataset |
	Test Error	Test Error
K_{80}	0.639 (0.03)	0.309 (0.06)
G_{80}	0.196 (0.03)	0.271 (0.06)
K_{50}	0.644 (0.02)	0.311 (0.03)
G_{50}	0.195 (0.02)	0.251 (0.04)
K_{20}	0.648 (0.01)	0.308 (0.02)
G_{20}	0.256 (0.04)	0.218 (0.03)

Table 7: Parzen window estimate for Breast cancer (Gaussian kernel sigma = 6) and Ionosphere datasets (Linear kernel). Quoted values are for the mean (and standard deviation) of the estimated generalisation error for the Parzen window estimate over twenty runs.

8.4.2. Kernel Selection and Combination

The concentration of the alignment can be directly used for tuning a kernel family to the particular task, or for selecting a kernel from a set, with no need for training. The probability that the level of alignment observed on the training set will be out by more than $\hat{\epsilon}$ from its expectation for one of the kernels is bounded by δ, where $\hat{\epsilon}$ is given by equation (8.39) for

$$\epsilon = \sqrt{\frac{8}{m}\left(ln\,|N|+ln\frac{2}{\delta}\right)}, \tag{8.39}$$

where $|N|$ is the size of the set from which the kernel has been chosen. One of the main consequences of the definition of kernel alignment is in providing a practical criterion for combining kernels. We will justify the intuitively appealing idea that two kernels with a certain alignment with a target that are not aligned to each other, will give rise to a more aligned kernel combination. In particular we have that

$$A_{k_1+k_2}(y) = \frac{\langle k_1 + k_2, y \otimes y \rangle_P}{\langle k_1 + k_2 \rangle_P}$$

$$= \frac{\langle k_1, y \otimes y \rangle_P}{\langle k_1 + k_2 \rangle_P} + \frac{\langle k_2, y \otimes y \rangle_P}{\langle k_1 + k_2 \rangle_P}$$

$$= A_{k_1}(y)\frac{\langle k_1 \rangle_P}{\langle k_1 + k_2 \rangle_P} + A_{k_2}(y)\frac{\langle k_2 \rangle_P}{\langle k_1 + k_2 \rangle_P}$$

This shows that if two kernels with equal alignment to a given target y are also completely aligned to each other, then $\|k_1 + k_2\|_P = \|k_1\|_P + \|k_2\|_P$ and the alignment remains of the combined kernel remains the same. If on the other hand the kernels are not completely aligned, then

$$\frac{\|k_1\|_P}{\|k_1 + k_2\|_P} + \frac{\|k_2\|_P}{\|k_1 + k_2\|_P} > 1, \tag{8.40}$$

and so the alignment of the combined kernel is correspondingly increased. Intuitively, this can be illustrated if we assume

$$k_1 = y \otimes y + \sum_{i=1}^{n} \lambda_i y^i \otimes y^i \tag{8.41}$$

and

$$k_2 = y \otimes y + \sum_{i=1}^{n} \mu_i y^i \otimes y^i. \tag{8.42}$$

In fact, the fact that they are both fairly aligned with the target means that they have a strong component corresponding to it, and the fact that they are not aligned to each other means that the other components are different, say $\mu_i \lambda_i = 0$ for all i. Averaging them will hence emphasize the part aligned with the target, and de-emphasize the other components. The resulting kernel will be less general, and more focused on the specific learning problem, in particular if the y and y^i are orthogonal we have that

$$A_{k_1}(y) = \frac{1}{\sqrt{1 + \sum_i \lambda_i^2}}$$

$$A_{k_2}(y) = \frac{1}{\sqrt{1 + \sum_i \mu_i^2}}$$

while

$$A_{k_1 + k_2}(y) = \frac{2}{\sqrt{4 + \sum_i (\lambda_i + \mu_i)^2}}$$

$$= \frac{1}{\sqrt{1 + 0.25 \sum_i \lambda_i^2 + 0.25 \sum_i \mu_i^2}}$$

which if $\sum_i \lambda_i^2 \approx \sum_i \mu_i^2$ gives

$$A_{k_1 + k_2}(y) \approx \frac{1}{\sqrt{1 + 0.5 \sum_i \lambda_i^2}}$$

$$\approx A_{k_1}(y) \sqrt{\frac{2}{1 + A_{k_1}(y)^2}}.$$

One can also set up the general problem of optimizing the coefficients of a combination like that $k = \sum \lambda_i k_i$, where one can choose very different kernels k_i. This is the subject of ongoing research.

8.4.3. Optimization over Combination of Kernels

We now consider a general linear combination of kernels

$$K(\alpha) = \sum_{k=1}^{T} \alpha_k K_k,$$

and study the problem of choosing α to optimize the alignment of $K(\alpha)$ to some given target vector y. Define $A(\alpha)$ as

$$A(\alpha) = A(S, K(\alpha), yy') = \frac{\sum_k \alpha_k y' K_k y}{m \sqrt{\sum_{kl} \alpha_k \alpha_l \langle K_k, K_l \rangle_F}}.$$

Hence, to maximize the alignment quantity we maximize $A(\alpha)$ subject to the constraint $\alpha_i \geq 0$. This constraint can be weakened in some cases if we are prepared to solve a semi-definite program [14]. For the purposes of this paper we will see that it will be sufficient to restrict ourselves to the case $\alpha_i \geq 0$. Hence, we must solve

$$\max_{\alpha} A(\alpha) \text{ subject to } \alpha_i \geq 0,$$

which is equivalent to

$$\max_{\alpha} \sum_k \alpha_k y' K_k y \text{ subject to } \sum_{kl} \alpha_k \alpha_l \langle K_k, K_l \rangle_F = C, \alpha_i \geq 0.$$

Applying the Lagrange multiplier method, we obtain,

$$\max_{\alpha} \sum_k \alpha_k y' K_k y - \lambda \left(\sum_{kl} \alpha_k \alpha_l \langle K_k, K_l \rangle_F - C \right) \text{ subject to } \alpha_i \geq 0.$$

Varying C leads to different values of λ. Since the alignment is invariant to rescaling α, we can choose $\lambda = 1$, fixing some value for the denominator and minimizing the numerator. Hence, we obtain the optimization problem

$$\max_{\alpha} \sum_k \alpha_k y' K_k y - \sum_{kl} \alpha_k \alpha_l \langle K_k, K_l \rangle_F \text{ subject to } \alpha_i \geq 0.$$

The second problem that arises is that if we do not constrain $\| \alpha \|$, the kernel can overfit its alignment to the training set, making its 'generalization' alignment on the test set poor. Hence, in exactly the same way as constraining the norm in for example Ridge Regression prevents overfitting, constraining $\| \alpha \|$ prevents over-aligning. Including this and using the Lagrange multiplier method again, we obtain

$$\max_{\alpha} \sum_k \alpha_k y' K_k y - \sum_{kl} \alpha_k \alpha_l \langle K_k, K_l \rangle_F - \lambda \sum_k \alpha_k^2$$
$$= \sum_k \alpha_k y' K_k y - \sum_{kl} \alpha_k \alpha_l \left(\langle K_k, K_l \rangle_F + \lambda \delta_{kl} \right)$$

subject to $\alpha_i \geq 0$.

The resulting optimization has a very similar form to that of the Support Vector Machine optimization problem. The first linear term is a sum of positive factors of the α_k, that is $y' K_k y \geq 0$, in place of the values 1 in the SVM case. The second part is a quadratic function, which as in the SVM case, is convex since

$$\sum_{kl} u_k u_l (\langle K_k, K_l \rangle_F + \lambda \delta_{kl}) = \left\langle \sum_k u_k K_k, \sum_l u_l K_l \right\rangle + \sum_k u_k u_k$$
$$= \| \sum_k u_k K_k \|_F^2 + \| u \|^2 \geq 0.$$

Note also that by Proposition 2 the entries in the Hessian are all positive. Hence, we can solve for α using the standard quadratic programming methods used in the SVM optimization. Furthermore, we can expect the α vector to be sparse, that is that only a subset of the kernels will be included in the optimal combination.

8.4.4 Scalable Kernel Alignment Optimization

The approach described in Section 8.4.1 for optimizing alignment required the eigendecomposition of the complete kernel matrix which can be computationally prohibitive especially for large kernel matrices. In this section we propose a general method for optimizing alignment over a linear combination of kernels that is suitable for large kernel matrices. This approach gives rise to both transductive and a novel inductive algorithms based on the Incomplete Cholesky factorization of the kernel matrix. The Incomplete Cholesky factorization is equivalent to performing a Gram-Schmidt orthogonalization of the training points in the feature space.

The results show that optimizing the alignment of the projection of the data into a 1-dimensional subspace is equivalent to performing Ridge Regression (RR). Furthermore, the alignment of the full kernel matrix lower bounds its projected value. Together these results show that optimizing the alignment followed by a Ridge Regression optimization gives a well founded model selection strategy. Experimental results are presented in Section 8.4.6. The approach adopted in this paper is related to that presented by [14]. Here we optimize the alignment, and subsequently its projection in a ridge regression style algorithm. In contrast, [14] consider an alternative optimization approach based on optimizing the margin using semi-definite programming (SDP).

8.4.4.1 Gram-Schmidt Optimization

In kernel based methods, large datasets pose significant problems since the number of basis functions required for an optimal solution can equal the number of data samples. A number of methods have been proposed for obtaining a low rank matrices such that the Frobenius norm is minimized [23]. In this work, an approximation strategy, based on the Gram-Schmidt decomposition in the feature space is considered. This algorithm is equivalent to the incomplete Cholesky decomposition of the kernel matrix used by [1] for kernel independent components analysis (kernel ICA). The projection is built up as the span of a subset of (the projections of) a set of k training examples. These are selected by performing a Gram-Schmidt orthogonalization of the training vectors in the feature space. Hence, once a vector is selected the remaining training points are transformed to become orthogonal to it. The next vector selected is the one with the largest residual norm. The whole transformation is performed in the feature space using the kernel mapping to represent the vectors obtained. The method is parameterized by the number of dimensions T selected. Figure 1 shows the pseudo-code for the Gram-Schmidt/incomplete Cholesky factorization algorithm. If we now create the vectors v_k, $k = 1,...,T$, by setting

$$v_{ki} = \frac{feat[i,k]}{\sqrt{\sum_{i'} feat[i',k]^2}},$$

(where $feat$ is defined in the pseudo-code) then we can express the approximate reconstruction of K as $K \approx \sum_{k=1}^{T} d_k v_k v'_k$, where $d_k = \sum_{i'} feat[i',k]^2$, since

$$K_{ij} \approx \sum_{k=1}^{T} feat[i,k] feat[j,k] = \sum_{k=1}^{T} v_{ki} v_{kj} \sum_{i'} feat[i',k]^2.$$

Hence, the Gram-Schmidt decomposition of K returns the matrix V and a diagonal matrix D with diagonal entries d_k. This approximation is constructed by choosing a sequence of exemplar training examples that most completely span the space and projecting the data into those directions.

Require: A kernel k, training set $\{(d_1, y_1), ..., (d_n, y_n)\}$ and number T [0]

    for $i = 1$ to n **do**

        $norm[i] = k(d_i, d_i)$;

    end for

    for $j = 1$ to T **do**

        $i_j = argmax_i(norm[i])$;

        $index[j] = i_j$;

        $size[j] = \sqrt{norm[i_j]}$;

        for $i = 1$ to n **do**

        $feat[i, j] = \frac{\left(k(d_i, d_{i_j}) - \sum_{t=1}^{j-1} feat[i,t] * feat[i_j, t]\right)}{size[j]}$;

        $norm[i] = norm[i] - feat[i, j] * feat[i, j]$;

        end for

    end for

return $feat[i, j]$ as the j-th feature of input i;

Figure 1: The Gram-Schmidt Algorithm

It therefore suggests that we explore the linear combination of kernels, $K(\alpha) = \sum_{k=1}^{T} \alpha_k v_k v'_k$, and apply the methods of optimization developed in the previous section.

Note that while in combining a general set of kernels it may happen that a combination with negative coefficients is still positive semi-definite, we show that here for $K(\alpha)$ to be positive semi-definite we must have $\alpha_k \geq 0$ for all k. This follows from the fact that if the order in which the vectors are orthogonalized is given by $i_1, i_2, ..., i_T$, then the entries $v_{ki_\ell} = 0$, for $\ell < k$ and $v_{ki_k} \neq 0$ (assuming the k-th feature vector has non-zero residual). Hence, the matrix V with the vectors v_k as rows has rank T. Now assume $\alpha_{\bar{k}} < 0$. Since V has full row rank, there is a vector u such that $Vu = e_{\bar{k}}$, the $\bar{k}$-th unit vector. Now

$$u'K(\alpha)u = \sum_k \alpha_k u'v_k v_k u = \alpha_{\bar{k}} < 0.$$

Hence, we obtain the optimally aligned linear combination by solving the optimization,

$$\max_{\alpha} \sum_k \alpha_k (y'v_k)^2 - \sum_{kl} \alpha_k \alpha_l \left((v'_k v_l)^2 + \lambda \delta_{kl} \right) \text{ subject to } \alpha_i \geq 0.$$

Though we have described the algorithm working on training and test datasets at the same time, in what is effectively a transductive setting, there is no reason why the computation of the features for the unlabelled data cannot be delayed provided the relevant information from the training stage is stored. Hence, this algorithm can also be used in an inductive setting, the only difference being that the testing data cannot be chosen in the orthogonalization procedure.

8.4.5 Non-margin based Gram-Schmidt Optimization

We state the following two theorems (proofs omitted) that relate the optimization of the projected alignment to the Ridge Regression optimization:

$$\min_w L(w) = \lambda \langle w, w \rangle + \sum_{i=1}^m \left(\langle w, \mathbf{x}_i \rangle - y_i \right)^2. \tag{8.43}$$

Theorem 13 *Let X be a feature/example matrix expressed in a possibly kernel-defined feature space. The solution of the optimization*

$$argmax_{w:\|w\|\leq 1} A(S, X'ww'X, yy')$$

gives the weight vector that solves the Ridge Regression problem (43) with the regularization parameter $\lambda = 0$.

Theorem 14 *Let X be a feature/example matrix expressed in a possibly kernel-defined feature space. The solution of the optimization*

$$w_* = argmax_{w:\|w\|\leq 1} A(S, X'ww'X, yy')$$

satisfies

$$A(S, X'w_*w_*X, yy') \geq A(S, X'X, yy').$$

Together the theorems show that optimizing the alignment decreases a lower bound on the objective of the Ridge Regression optimization. This suggests that optimizing the alignment will lead to better generalization performance of two norm error bound classifiers.

8.4.6 Experiments

To demonstrate the performance of the algorithms presented we have used two binary classification datasets. The ionosphere dataset[4] which contains 34 inputs, a single binary output and 351 datapoints. The Wisconsin breast cancer dataset contains nine integer valued inputs, a single binary output (benign or malignant tumours) and 699 datapoints. Three learning algorithms were implemented. A Parzen window estimator, a support vector classifier (SVC) and a kernel ridge regressor (treating the binary labels +1, −1 as real targets). A 10-fold cross validation procedure was used to find the optimal values for the capacity control parameter 'C'. Having selected the optimal value of C, the SVC was re-trained ten times using ten random data splits. A similar procedure was used to find the optimal Ridge Regression parameter λ. A linear kernel was used for training for both of the datasets. The K matrices are before adaptation, while the G matrices are after optimization using the transductive and inductive alignment algorithms. The index represents the percentage of training points.

	TRAIN ALIGN	TEST ALIGN	PW ERROR	SVC ERROR	RR ERROR
K_{80}	0.244 (0.012)	0.258 (0.046)	0.320 (0.052)	0.279 (0.042)	0.240 (0.048)
G_{80}	0.342 (0.012)	0.350 (0.089)	0.214 (0.030)	0.206 (0.024)	0.207 (0.039)
K_{50}	0.224 (0.029)	0.272 (0.036)	0.297 (0.036)	0.262 (0.033)	0.244 (0.039)
G_{50}	0.360 (0.028)	0.309 (0.034)	0.209 (0.016)	0.205 (0.017)	0.203 (0.015)
K_{20}	0.254 (0.032)	0.245 (0.008)	0.310 (0.013)	0.281 (0.020)	0.279 (0.016)
G_{20}	0.316 (0.031)	0.286 (0.016)	0.231 (0.021)	0.223 (0.024)	0.222 (0.020)

Table 8 Ionosphere dataset – alignment values, SVC, Parzen window(PW) and ridge Regression(RR) error for a linear kernel over 10 runs using transductive Gram-Schmidt.

Table 8 presents the results obtained from the transductive alignment algorithm applied to the ionosphere dataset. As observed from these results the alignment for all of the training portions considered increases for the adapted matrix G over that of the original matrix K. A similar effect is observed on the test set for the adapted matrix G. The generalization error values for all three learning algorithms considered shows a reduction for

---

[4]available from the UCI data repository

the adapted matrix with increasing training and test set alignment. It is interesting to note the good generalization performance of the ridge regression algorithm for both the original kernel matrix K and the adapted matrix G compared to the support vector classifier and Parzen window estimator.

	TRAIN ALIGN	TEST ALIGN	PW ERROR	SVC ERROR	RR ERROR
K_{80}	0.243 (0.013)	0.261 (0.053)	0.323 (0.047)	0.281 (0.037)	0.253 (0.033)
G_{80}	0.311 (0.013)	0.285 (0.052)	0.251 (0.053)	0.230 (0.046)	0.230 (0.040)
K_{50}	0.227 (0.033)	0.270 (0.040)	0.295 (0.034)	0.259 (0.032)	0.248 (0.034)
G_{50}	0.303 (0.040)	0.260 (0.039)	0.239 (0.041)	0.223 (0.044)	0.204 (0.031)
K_{20}	0.266 (0.066)	0.243 (0.016)	0.314 (0.017)	0.286 (0.014)	0.275 (0.035)
G_{20}	0.375 (0.064)	0.109 (0.017)	0.259 (0.058)	0.260 (0.055)	0.243 (0.066)

Table 9 Ionosphere dataset - alignment values, SVC and Parzen window (PW) and Ridge Regression (RR) error for a linear kernel over 10 runs using inductive Gram-Schmidt.

Table 9 presents the results obtained from the inductive alignment algorithm using the Ionosphere dataset. From the values quoted in the table, a similar trend to that for the transductive algorithm is observed. The alignment on both the training and the test datasets increases for the adapted matrix G across all of the data partitions. This is accompanied by a corresponding decrease in the generalization error for the learning algorithms considered. Comparing the generalization error estimates for the transductive and inductive alignment algorithms, the transductive version outperforms the inductive alignment algorithm.

	TRAIN ALIGN	TEST ALIGN	PW ERROR	SVC ERROR	RR ERROR
K_{80}	0.112 (0.007)	0.137 (0.031)	0.336 (0.024)	0.222 (0.034)	0.336 (0.023)
G_{80}	0.251 (0.006)	0.294 (0.037)	0.247 (0.030)	0.131 (0.032)	0.244 (0.032)
K_{50}	0.120 (0.019)	0.115 (0.020)	0.353 (0.017)	0.250 (0.021)	0.356 (0.017)
G_{50}	0.269 (0.019)	0.245 (0.023)	0.262 (0.021)	0.139 (0.019)	0.259 (0.021)
K_{20}	0.116 (0.040)	0.117 (0.010)	0.349 (0.008)	0.242 (0.012)	0.349 (0.008)
G_{20}	0.259 (0.039)	0.242 (0.010)	0.267 (0.022)	0.146 (0.009)	0.266 (0.021)

Table 10 Breast dataset – alignment values, SVC and Parzen window (PW) and Ridge Regression (RR) error for a linear kernel over 10 runs using inductive Gram-Schmidt.

Table 10 presents the results obtained from the transductive alignment algorithm applied to the breast cancer dataset. A similar effect to that observed for the Ionosphere dataset (table 8) is seen for this dataset. The alignment for all of the training portions considered increases for the adapted matrix G over that of the original matrix K. A similar effect is observed on the test set for the adapted matrix G. The generalization error values for all three learning algorithms considered shows a reduction for the adapted matrix with increasing training and test set alignment. It is interesting to note the good generalization performance of the simple Parzen window estimator for both the original kernel matrix K and the adapted matrix G compared to the support vector classifier and ridge regression algorithm.

	TRAIN ALIGN	TEST ALIGN	SVC ERROR	PW ERROR	RR ERROR
K_{80}	0.079 (0.008)	0.080 (0.034)	0.239 (0.107)	0.653 (0.045)	0.355 (0.027)
G_{80}	0.311 (0.008)	0.321 (0.031)	0.115 (0.023)	0.133 (0.027)	0.167 (0.039)
K_{50}	0.089 (0.018)	0.070 (0.016)	0.240 (0.156)	0.663 (0.021)	0.361 (0.018)
G_{50}	0.312 (0.017)	0.308 (0.016)	0.115 (0.009)	0.134 (0.013)	0.204 (0.058)
K_{20}	0.081 (0.035)	0.079 (0.010)	0.188 (0.087)	0.649 (0.011)	0.347 (0.008)
G_{20}	0.328 (0.034)	0.295 (0.009)	0.139 (0.014)	0.142 (0.012)	0.254 (0.069)

Table 11 Breast dataset – alignment values, SVC and Parzen window (PW) and Ridge Regression (RR) error for a linear kernel over 10 runs using inductive Gram-Schmidt.

Table 11 presents the results obtained from the inductive alignment algorithm using breast cancer dataset. From the values quoted in the table, a similar trend to that for the transductive algorithm is observed. The alignment on both the training and the test datasets increases for the adapted matrix G across all of the data partitions. This is accompanied by a corresponding decrease in the generalization error for the learning algorithms considered.

8.5 Clustering by Maximal Alignment

In this and the next subsection we will introduce two related methods for clustering (they can also be extended to the case of transduction). We introduce a new quantity, that depends on the kernel and on the (unlabeled) dataset. It depends on the relation between kernel and input distribution. We call absolute alignment of a kernel to a set of data points the maximum of the alignment over all possible labelings, and we call the corresponding labeling 'optimally aligned' or optimal.

$$\hat{A}^*_k = \max_{y \in \{-1,1\}^m} \hat{A}_k(y) = \max_{y \in \{-1,1\}^m} \frac{\langle K, yy' \rangle_F}{\sqrt{\langle K, K \rangle_F \langle yy', yy' \rangle_F}} \qquad (8.44)$$

Of course since the solution vector is $y \in \{-1, +1\}^m$ we obtain a combinatorial optimization problem. We can relax this constraint by simply asking that $\| y \|^2_2 = m$ (the exact problem is of course a special case of this setting). Then after solving the relaxed problem, we can obtain an approximate discrete solution by choosing a suitable threshold. In order to solve this problem we will use the following characterization of the spectrum of a symmetric matrix.

Theorem 15 (Courant-Fischer Minimax Theorem)

If $M \in \mathbb{R}^{m \times m}$ is symmetric, then

$$\lambda_k(M) = \max_{dim(T)=k} \min_{0 \neq v \in T} \frac{v' M v}{v' v} = \min_{dim(T)=m-k+1} \max_{0 \neq v \in T} \frac{v' M v}{v' v}, \qquad (8.45)$$

for $k = 1, ..., m$.

By a simple change of variables, it is possible to see that the approximated maximum-alignment problem is solved by the first eigenvector, and hence that the absolute alignment is upper bounded by the first eigenvalue. For the maximum eigenvalue, only the *max* matters. So

$$\lambda_{max} = \max_{0 \neq v \in \mathbb{R}^m} \frac{v' K v}{v' v} \qquad (8.46)$$

One can now transform this vector into a vector in $\{-1, +1\}^m$ by choosing the threshold θ that gives maximum alignment of $y = sign(v^{max} - \theta)$. By definition, the value of alignment $A(y)$ obtained by this solution will be a lower bound of the optimal alignment, that hence we have

$$\hat{A}(y) \leq \hat{A}^* \leq \frac{1}{\sqrt{\langle K, K \rangle_F}} \lambda_{max} = \frac{\lambda_{max}}{\| K \|_F}. \qquad (8.47)$$

One can hence estimate the quality of a dichotomy by comparing its value with the upper bound. The absolute alignment tells us how specialized a kernel is on a given dataset: the higher this quantity, the more committed to a specific dichotomy is the kernel. This leads to a simple algorithm for transduction or unsupervised learning: given a set of points and a kernel, fill the kernel matrix. If some labels are available, insert in the corresponding entry $K(x_i, x_j) = K_{\max}$ if $y_i = y_j$ and $K_{\min}$ otherwise. Choose the kernel parameters so as to optimize $\lambda_{\max} / \|K\|$. Then find the first eigenvector, choose a threshold to maximize the alignment and output the corresponding y. The cost to the alignment of changing a label y_i is

$$2\frac{\sum_j y_j k(x_i, x_j)}{\|k\|}, \tag{8.48}$$

so that if a point is isolated from the others, or if it is equally close to the two different classes, then changing its label will have only a very small effect. Vice versa, labels in strongly clustered points clearly contribute to the overall cost and changing their label will alter the alignment significantly.

If many entries of the first eigenvector are close to 0, one could proceed in iterations, by choosing those entries whose labels are more certain, and treating them as given labels, so modifying the kernel matrix as defined above. Then the remaining labels can be optimized again. The first eigenvector can be calculated in many ways, for example the Lanczos procedure, which is already effective for large datasets. Search engines like Google are based on estimating the first eigenvector of a matrix with dimensionality more than 10^9, so for very large datasets it may be possible to develop approximate techniques.

Figure 1 shows the actual alignment, test error and upper bound on absolute alignment as a function of the threshold for breast cancer data, for the above algorithm. Notice that where actual alignment and upper bound on alignment get closest, we have confidence that we have partitioned our data well, and in fact the test accuracy is also maximized. Notice also that the choice of the threshold corresponds to maintaining the correct proportion between positives and negatives. This suggests another possible threshold selection strategy, based on the availability of enough labeled points to give a good estimate of the proportion of positive points in the dataset.

In Figure 2 we plotted the upper and lower bounds for the alignment, together with the error, for unsupervised classification of cancer data, as a function of the threshold, for the first eigenvector. We see that where the alignment is optimal, also the error is optimal. The first eigenvector can be calculated in many ways, for example the Lanczos procedure, which is already effective for large sparse datasets. Search engines like Google are based on estimating the first eigenvector of a matrix with dimensionality more than 10^9, so for very large datasets there are approximation techniques.

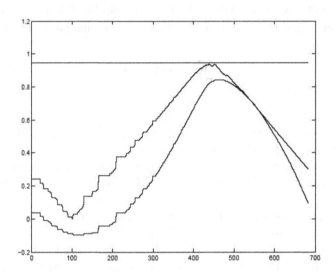

Figure 2 Plot for cancer data of $\lambda_{max}/\sqrt{\langle K,K\rangle_F}$, of $A(y)$ for $y = sign(v_{max})$, and of the error (curve in the middle).

We applied the procedure outlined above to two datasets from the UCI repository. We preprocessed the data by normalizing the input vectors in the kernel defined feature space and then centering them by shifting the origin (of the feature space) to their centre of gravity. This can be achieved by the following transformation of the kernel matrix,

$$K \longleftarrow K - m^{-1}jg' - m^{-1}gj' + m^{-2}j'KjJ \qquad (8.49)$$

where j is the all ones vector, J the all ones matrix and g the vector of row sums of K.

The first experiment applied the unsupervised technique to the Breast Cancer data with a linear kernel. Figure 3(a) shows the alignment of the different eigenvectors with the labels. The highest alignment is shown by the last eigenvector corresponding to the largest eigenvalue. For each value θ_i of the threshold Figure 3(b) shows the upper bound of $\lambda_{max}/\|K\|_F$ (straight line), the alignment $\hat{A}(S,k,y)$ for $y = sign(v^{max} - \theta_i)$ (bottom curve), and the accuracy of y (middle curve). Notice that where actual alignment and upper bound on alignment get closest, we have confidence that we have partitioned our data well, and in fact the accuracy is also maximized.

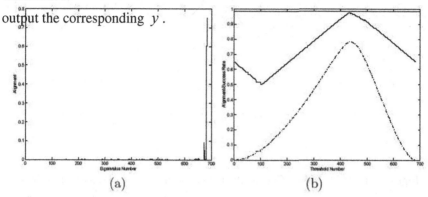

(a)   (b)

Figure 3(a) Plot of alignment of the different eigenvectors with the labels ordered by increasing eigenvalue. **(b)** Plot for Breast Cancer data (linear kernel) of $\lambda_{max}/\|K\|_F$ (straight line), $\hat{A}(S,k,y)$ for $y = sign(v^{max} - \theta_i)$ (bottom curve), and the accuracy of y (middle curve) against threshold number i.

Notice also that the choice of the threshold corresponds to maintaining the correct proportion between positives and negatives. This suggests another possible threshold selection strategy, based on the availability of enough labeled points to give a good estimate of the proportion of positive points in the dataset. This is one way label information can be used to choose the threshold. At the end of the experiments we will describe another 'transduction' method. It is a measure of how naturally the data separates that this procedure is able to optimize the split with an accuracy of approximately 97.29% by choosing the threshold that maximizes the alignment (threshold number 435) but *without making any use of the labels*.

In Figure 4a we present the same results for the Gaussian kernel ($\sigma = 6$). In this case the accuracy obtained by optimizing the alignment (threshold number 316) of the resulting dichotomy is less impressive being only about 79.65%. Finally, Figure 4b shows the same results for the Ionosphere dataset. Here the accuracy of the split that optimizes the alignment (threshold number 158) is approximately 71.37%. We can also use the overall approach to adapt the kernel to the data. For example we can choose the kernel parameters so as to optimize $\lambda_{max}/\|K\|_F$. Then find the first eigenvector, choose a threshold to maximize the alignment and output the corresponding y.

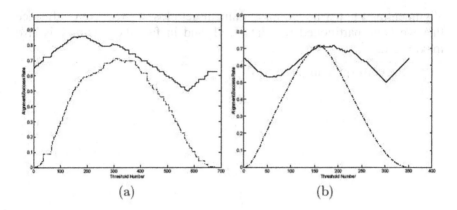

(a)   (b)

Figure 4 Plot for Breast Cancer data (Gaussian kernel) **(a)** and Ionosphere data (linear kernel) **(b)** of $\lambda_{max}/\|K\|_F$ (straight line), $\hat{A}(S,k,y)$ for $y = sign(v^{max} - \theta_i)$ (bottom curve), and the accuracy of y (middle curve) against threshold number i.

The cost to the alignment of changing a label y_i is $2\sum_j y_j k(x_i,x_j)/\|K\|_F$, so that if a point is isolated from the others, or if it is equally close to the two different classes, then changing its label will have only a very small effect. On the other hand, labels in strongly clustered points clearly contribute to the overall cost and changing their label will alter the alignment significantly.

The method we have described can be viewed as projecting the data into a 1-dimensional space and finding a threshold. The projection also implicitly sorts the data so that points of the same class are nearby in the ordering. We discuss the problem in the 2-class case. We consider embedding the set into the real line, so as to satisfy a clustering criterion. The resulting Kernel matrix should appear as a block diagonal matrix. This problem has been addressed in the case of information retrieval in [4], and also applied to assembling sequences of DNA. In those cases, the eigenvectors of the Laplacian have been used, and the approach is called the Fiedler ordering. Although the Fiedler ordering could be used here as well, we present here a variation based on the simple kernel matrix

Let the coordinate of the point x_i on the real line be $v(i)$. Consider the cost function $\sum_{ij} v(i)v(j)K(i,j)$. It is maximized when points with high similarity have the same sign and high absolute value, and when points with different sign have low similarity. The choice of coordinates v that optimizes this cost is the first eigenvector, and hence by sorting the data according to the value of their entry in this eigenvector one can hope to find a good permutation, that renders the kernel matrix block diagonal. Figure 5 shows the results of this heuristic applied to the Breast cancer dataset. The grey level indicates the size of the kernel entry. The figure on the left is for the unsorted data, while that on the right shows the same plot after sorting. The sorted figure clearly shows the effectiveness of the method.

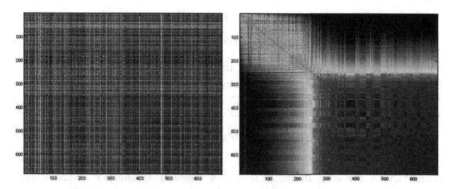

Figure 5 Gram matrix for cancer data, before and after permutation of data according to sorting order of first eigenvector of K

8.5.1. Clustering by Minimizing the Cut-Cost

Another measure of separation between classes is the average separation between two points in different classes, again normalized by the matrix norm.

Definition 16. *Cut Cost.* *The cut cost of a clustering is defined as*

$$C(S,k,y) = \frac{\sum_{ij:y_i \neq y_j} k(x_i, x_j)}{m \, \| K \|_F}. \tag{8.50}$$

This quantity is motivated by a graph theoretic concept. If we consider the Kernel matrix as the adjacency matrix of a fully connected weighted graph whose nodes are the data points, the cost of partitioning a graph is given by the total weight of the edges that one needs to cut or remove, and is exactly the numerator of the 'cut cost'. Notice also the relation between alignment and cutcost:

$$\hat{A}(S,k,y) = \frac{\sum_{ij} k(x_i, x_j) - 2C(S,k,y)}{m\sqrt{\langle K,K \rangle_F}} = T(S,k) - 2C(S,k,y), \quad (8.51)$$

where $T(S,k) = \hat{A}(S,k,j)$, for j the all ones vector. Notice that for normalized kernels $\|\phi(x_i) - \phi(x_j)\|^2 = 2 - 2K(x_i, x_j)$, and with a fixed K choosing y to minimize $\sum_{y_i \neq y_j} K_{ij}$ is equivalent to maximizing $\sum_{y_i = y_j} K_{ij}$, which is in turn equivalent to minimizing the sum of the squared distances between all couples of points in the same class $\sum_{y_i = y_j} \|\phi(x_i) - \phi(x_j)\|^2$. Furthermore minimising this quantity is equivalent to minimising the sum of the average square distances of points from their class means since $\sum_{y_i = C} \|\phi(x_i) - \mu(C)\|^2 = n_C - \frac{1}{n_C}\sum_{y_i = C = y_j} K_{ij}$, where we denote by $\mu(C) = \frac{1}{n_C}\sum_i \phi(x_i)$. So this approach directly aims at finding clusters that have minimal 'scatter' around their mean. Among other appealing properties of the alignment, is that this quantity is sharply concentrated around its mean, as proven earlier in this paper and also in [8]. This shows that the expected alignment can be reliably estimated from its empirical estimate $\hat{A}(S)$. As the cut cost can be expressed as the difference of two alignments

$$C(S,k,y) = 0.5(T(S,k) - \hat{A}(S,k,y)), \quad (8.52)$$

it will be similarly concentrated around its expected value.

This quantity can be approximately minimized by the eigenvectors of a matrix, called the Laplacian, derived from the kernel matrix, defined as $L = D - K$, where $K_{ij} = k(x_i, x_j)$ and $D = diag(d_1, ..., d_m)$ with $d_i = \sum_{j=1}^m k(x_i, x_j)$. The Laplacian matrix is positive semi-definite with smallest eigenvalue $\lambda_1 = 0$, and corresponding eigenvector with equal entries. The eigenvector corresponding to the next smallest eigenvalue λ_2 approximately gives the best *balanced* split, since it enforces the condition that the weight of positives is the same as the weight of negatives, by being orthogonal to the constant smallest eigenvector.

The above definitions imply the following two equations:

$$\sum_{y_i=y_j} K_{ij} - \sum_{y_i \neq y_j} K_{ij} = \sum_{ij} y_i y_j K_{ij} = y'Ky$$

$$\sum_{y_i=y_j} K_{ij} + \sum_{y_i \neq y_j} K_{ij} = \sum_{ij} K_{ij}$$

and so

$$\sum_{y_i \neq y_j} K_{ij} = C(y) = \frac{1}{2}\left(\sum_{i,j} K_{ij} - y'Ky\right) = -\frac{1}{2}y'Ly.$$

One would like to find $y \in \{-1,+1\}^m$ so as to minimize the cut cost, but this problem is NP-hard. We can impose a slightly looser constraint on y:
$y \in R^m$, $\sum_i y_i^2 = m$, $\sum_i y_i = 0$. We therefore have the problem

$$\min C(y)$$

subject to $\quad y \in \mathbb{R}^m, \sum_i y_i^2 = m, \sum_i y_i = 0.$

This is, however, identical to the second eigenvalue optimization

$$\max_{y \perp 1} \frac{-y'Ly}{2y'y} \tag{8.53}$$

Summarizing:

$$\min_{\sum_i y_i^2 = m, \sum_i y_i = 0} C(y) = \frac{m}{2}\lambda_2. \tag{8.54}$$

So the eigenvector corresponding to the second eigenvalue of the Laplacian can be used to obtain a good approximate split and the second eigenvalue of the Laplacian gives a lower bound on the cut-cost. One can now threshold the entries of the eigenvector in order to obtain a vector with -1 and $+1$ entries. The choice of the threshold could be discussed further, but for our purposes we will set it to 0. In Section 8.6.5 we will see a method for adapting the first eigenvector, making it similar to a chosen one. This method could be used to incorporate into the first eigenvector information about the relative size of the positive and negative class, so that the second eigenvector will provide the most aligned solution that has the chosen ration of positives / negatives.

Remark. It is also easy to see that the number of zero eigenvalues in L (and hence its rank) is an estimation of the number of clusters in the data.

Remark. One could also consider the transition matrix T of a random walk on the graph, where the probabilities are proportional to the weights. This would be obtained by removing the diagonal of the Gram matrix and normalizing each row. This quantity would give information about anomalous points. These points and further extensions of alignment are currently being considered by us.

We applied the procedure to the Breast cancer data with both linear and Gaussian kernels. The results are shown in Figure 6.

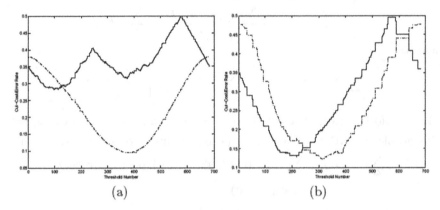

(a)   (b)

Figure 6 Plot for Breast Cancer data using (a) Linear kernel) and (b) Gaussian kernel of $C(S,k,y) - \lambda/(2\|K\|_F)$ (dashed curves), for $y = sign(v^{max} - \theta_i)$ and the error of y (solid curve) against threshold number i.

Now using the cut cost to select the best threshold for the linear kernel sets it at 378 with an accuracy of 67.86%, significantly worse than the results obtained by optimizing the alignment. With the Gaussian kernel, on the other hand, the method selects threshold 312 with an accuracy of 80.31%, a slight improvement over the results obtained with this kernel by optimizing the alignment.

So far we have presented algorithms that use unsupervised data. We now consider the situation where we are given a partially labelled dataset. This leads to a simple algorithm for transduction or semi-supervised learning. The idea that some labelled data might improve performance comes from observing Figure 6b, where the selection based on the cut-cost is clearly suboptimal. By incorporating some label information, it is hoped that we can obtain an improved threshold selection.

Let z be the vector containing the known labels and 0 elsewhere. Set $KP = K + C_0 zz'$, where C_0 is a positive constant parameter. We now use the original matrix K to generate the eigenvector, but the matrix KP when measuring the cut-cost of the classifications generated by different thresholds. Taking $C_0 = 1$ we performed 5 random selections of 20% of the data and obtained a mean success rate of 85.56% (standard deviation 0.67%) for the Breast cancer data with Gaussian kernel, a marked improvement over the 80.31% achieved with no label information.

8.6 Conclusions

The problem of assessing the quality of a kernel is central to the theory of kernel-machines, and deeply related to the problem of model/feature selection as a whole. Being able to quantify this property is an important step towards effective algorithms for kernel selection, combination and adaptation.

The quality of a kernel is necessarily not an absolute property, but relative to a given learning task. So any measure of kernel quality should actually measure the degree of agreement between a kernel and a labeling, or even between two kernels.

In this paper we proposed one such quantity, which we call Alignment, we discussed a number of theoretical properties (including a novel characterization of kernel functions in its terms, its statistical concentration under a random choice of data sample, and its connection with the generalization properties of certain algorithms) and we derived a number of algorithms based on the idea of optimizing the alignment between a kernel and the learning target, demonstrating that its simplicity makes this quantity a very versatile conceptual tool.

We gave a criterion to decide when two kernels can be combined into a better one, a practice recently used for hypertext (combining link and words information [7]) and bioinformatics (combining gene expression and philogenetic information [16]).

We demonstrated a general method for optimizing alignment over a linear combination of kernels. The approach we developed has been extended to give both transductive and inductive algorithms based on the Incomplete Cholesky factorization of the kernel matrix. The method is based upon the incomplete Cholesky factorization, which as we argue in the paper is equivalent to performing a Gram-Schmidt orthogonalization of the training points in the feature space.

The alignment optimization method adapts the feature space to increase its training set alignment. Regularization is required to ensure this alignment is also retained for the test set, and ensures that a sparse solution is obtained. In this paper we provided both theoretical and experimental evidence to show that improving the alignment leads to a reduction in generalization error of standard learning algorithms.

We gave also criteria for adapting kernels to a given target, both by fitting a parameterized kernel family and in a novel - non parametric fashion, where just the entries of the Gram matrix are adapted. Then we gave a series of methods for assigning the labels of a set of points so as to optimize the alignment with a given kernel. This leads to novel procedures for clustering and transduction.

Many of the ideas presented in this work have close relatives in previously or simultaneously appearing publications. Recent work from [12,2], explores the problem of embedding a certain matrix into an Euclidean space in order to derive inherent limitations of kernel based representations.

The use of spectral methods for clustering has been first introduced in graph theory [17]. Spectral Clustering algorithms have been recently discussed and analyzed [11], while Kernel PCA [20] and Latent Semantic Kernels [7] also are strongly related to the labeling methods discussed here. Other algorithms for clustering and transduction inspired by graph theory have also been recently put forward [5,22].

Theoretically, future work will include exploring the connections between high alignment and good generalization in larger classes of learning machines, and its relations with the luckiness framework [21], and the notion of stability.

References

[1]   F. Bach and M.I. Jordan. Kernel independent components analysis. Department of computer science, University of California, Berkeley, Berkeley, CA, 2001.

[2]   S. Ben-David, N. Eiron, and H.U. Simon. Non-embedability in Euclidean half spaces. In *NIPS Workshop on New Perspectives in Kernel Methods*, 2000.

[3]   K. P. Bennett and O. L. Mangasarian. Robust linear programming discrimination of two linearly inseparable sets. *Optimization Methods and Software*, 1:23-34, 1992.

[4]   M.W. Berry, B. Hendrickson, and P. Raghaven. Sparse matrix re-ordering schemes for browsing hypertext. In J. Renegar, M. Shub, and S. Smale, editors, *The Mathematics of Numerical Analysis*, pages 91-123. American Mathematical Society, 1996.

[5]   A. Blum and S. Chawla. Learning from labeled and unlabeled data using graph mincuts. In *International Conference on Machine Learning (ICML) 2001*, 2001.

[6]   B. Carl and I. Stephani. *Entropy, compactness and the approximation of operators*. Cambridge University Press, Cambridge, UK, 1990.

[7]   N. Cristianini, , J. Shawe-Taylor, and H. Lodhi. Latent semantic kernels. In *International Conference on Machine Learning (ICML 2000)*, 2000.

[8]   N. Cristianini, A. Elisseeff, J. Shawe-Taylor, and J. Kandola. On kernel target alignment. In *Neural Information Processing Systems 14 (NIPS 14)*, 2001.

[9]   N. Cristianini and J. Shawe-Taylor. *An introduction to Support Vector Machines*. Cambridge University Press, Cambridge, UK, 2000.

[10] L. Devroye, L. Gyorfi, and G. Lugosi. *A Probabilistic Theory of Pattern Recognition*. Number 31 in Applications of mathematics. Springer, New York, 1996.

[11] P. Drineas, R. Kannan, S. Vampala, A. Frieze, and V. Vinay. Clustering in large graphs and matrices. In *Proceedings of the 10th ACM-SIAM Symposium on Discrete Algorithms*, 1999.

[12] J. Forster, N. Schmitt, and H. U. Simon. Estimating the optimal margins of embeddings in euclidean half spaces. In *Submitted Computational Learning Theory (COLT) 2001*, 2001.

[13] R. Herbrich. *Learning Kernel Classifiers-Theory and Algorithms*. MIT Press, 2002.

[14] N. Lanckriet, N. Cristianini, P. Bartlett, L. El-Ghoui, and M.I Jordan. Learning the kernel matrix using semi-definite programming. In *International Conference on Machine Learning (ICML 2002)*, 2002.

[15] C. McDiarmid. On the method of bounded differences. In *Surveys in Combinatorics*, pages 148-188, 1989.

[16] P. Pavlidas, J. Weston, J. Cai, and W.N. Grundy. Gene functional classification from heterogeneous data. In *Proceedings of the Fifth International Conference on Computational Molecular Biology*, 2001.

[17] A. Pothen, H. Simon, and K. Liou. Partitioning sparse matrices with eigenvectors of graphs. *SIAM Journal Matrix Analysis*, 11(3):242-248, 1990.

[18] S. Saitoh. *Theory of Reproducing Kernels and its Applications*. Longman Scientific & Technical, Harlow, England, 1988.

[19] B. Scholkopf and A. Smola. *Learning With Kernels - Support Vector Machines, Regularization, Optimization and Beyond*. MIT Press, 2002.

[20] B. Scholkopf, A. Smola, and K. Muller. Kernel principal components analysis. In *Advances in Kernel Methods - Support Vector Learning*. MIT, 2000.

[21] J. Shawe-Taylor, P. L. Bartlett, R. C. Williamson, and M. Anthony. Structural risk minimization over data-dependent hierarchies. *IEEE Transactions on Information Theory*, 44(5) :1926-1940, 1998.

[22] J. Shi and J. Malik. Normalised cuts and image segmentation. In *In IEEE Conf. on Computer Vision and Pattern Recognition*, pages 731-737, 1997.

[23] A. Smola and B. Scholkopf. Sparse greedy matrix approximation for machine learning. In *In Proceedings International Conference on Machine Learning 2000*, 2000.

A   Proofs for the concentration of the Alignment

Proposition 17 *Let h and k be two kernels, let $S = \{x_1,..,x_m\}$ be a set of inputs and H (resp. K) the kernel matrix induced by h (resp. k) on S. Let $S^r = (S \setminus x_r) \cup x'_r$ and H^r (resp. K^r) the kernel matrix induced by h (resp. k) on S^r. Assume that $\|K\|_F$, $\|H\|_F$, $\|K^r\|_F$ and $\|H^r\|_F$ are all lower bounded by $a > 0$, then:*

$$\left| \hat{A}(k,h,S) - \hat{A}(k,h,S^r) \right| \le \frac{12}{ma^2} \tag{8.55}$$

Proof. S Let us rewrite:

$$\hat{A}(k,h,S^r) = \hat{A}(k+dk,h+dh,S) \tag{8.56}$$

where $(k+dk)(x,z)$ (resp. $(h+dh)(x,z)$) is the same as $k(x,z)$ (resp. $h(x,z)$) except that any value on x_r is replaced by its value on x'_r: $(k+dk)(x_r,z) = k(x'_r,z) \quad \forall z \in X$. We are thus left to bound $\left| \hat{A}(k,h,S) - \hat{A}(k+dk,h+dh,S) \right|$.

Let us first study $\hat{A}(k+dk,h,S) - \hat{A}(k+dk,h,S)$ for any h defined on S. We have:

$$\hat{A}(k+dk,h,S) - \hat{A}(k,h,S) = \frac{\langle k+dk,h \rangle}{\|k+dk\|\|h\|} - \frac{\langle k,h \rangle}{\|k\|\|h\|} \tag{8.57}$$

$$= \frac{\langle k+dk,h \rangle}{\|k+dk\|\|h\|} - \frac{\langle k+dk,h \rangle}{\|k\|\|h\|} + \frac{\langle k+dk,h \rangle}{\|k\|\|h\|} - \frac{\langle k,h \rangle}{\|k\|\|h\|} \tag{8.58}$$

$$= \frac{\langle k+dk,h \rangle}{\|h\|} \left(\frac{1}{\|k+dk\|} - \frac{1}{\|k\|} \right) + \frac{1}{\|k\|} \left\langle dk, \frac{h}{\|h\|} \right\rangle \tag{8.59}$$

$$= \underbrace{\frac{\langle k+dk,h \rangle}{\|k+dk\|\|h\|}}_{\le 1} \left(\frac{\|k\| - \|k+dk\|}{\|k\|} \right) + \frac{1}{\|k\|} \left\langle dk, \frac{h}{\|h\|} \right\rangle \tag{8.60}$$

But,

$$\left| \| k \| - \| k + dk \| \right| = \left| \frac{\| k \|^2 - \| k + dk \|^2}{\| k \| + \| k + dk \|} \right| \tag{8.61}$$

$$\leq \frac{1}{\| k \| + \| k + dk \|} \left(2\langle k, dk \rangle + \| dk \|^2 \right) \tag{8.62}$$

So that:

$$\hat{A}(k + dk, h, S) - \hat{A}(k, h, S)$$
$$\leq \frac{2\langle k, dk \rangle}{\| k \| (\| k \| + \| k + dk \|)} + \frac{\| dk \|^2}{\| k \| (\| k \| + \| k + dk \|)} + \left\langle \frac{dk}{\| k \|}, \frac{h}{\| h \|} \right\rangle \tag{8.63}$$

Now we look more closely at dk : we have

$$dk(x_r, x_r) = k(x'_r, x'_r) - k(x_r, x_r)$$

and, for $j \neq r$

$$dk(x_i, x_j) = dk(x_j, x_i) = \begin{cases} 0 & i \neq r \\ k(x'_r, x_j) - k(x_r, x_j) & \text{otherwise} \end{cases}$$

So that $\forall k$ uniformly bounded by 1:

$$\langle dk, k \rangle = \frac{1}{m^2} \sum_{j=1}^{m} dk(x_r, x_j) h(x_r, x_j) \tag{8.64}$$

$$\leq \frac{2}{m} \tag{8.65}$$

This implies that :

$$\frac{\| dk \|^2}{\| k \| (\| k \| + \| k + dk \|)} \leq \frac{2}{ma^2}$$

and,

$$\left\langle dk, \frac{h}{\| h \| \| k \|} \right\rangle \leq \frac{2}{ma^2} \tag{8.66}$$

$$\frac{2\langle k, dk \rangle}{\| k \| (\| k \| + \| k + dk \|)} \leq \frac{2}{ma^2} \tag{8.67}$$

Thus we have:

$$\hat{A}(k + dk, h, s) - \hat{A}(k, h, S) \leq \frac{6}{ma^2} \tag{8.68}$$

and:

$$\left| \hat{A}(k, h, S) - \hat{A}(k, h, S^r) \right| = \left| \hat{A}(k + dk, h + dh, S) - \hat{A}(k, h, S) \right| \tag{8.69}$$

$$= \left| \begin{array}{l} \hat{A}(k+dk,h+dh,S) - \hat{A}(k+dk,h,S) \\ + \hat{A}(k+dk,h,S) - \hat{A}(k,h,S) \end{array} \right| \tag{8.70}$$

$$\leq \frac{12}{ma^2} \tag{8.71}$$

Proposition 18 *Let S be a set of inputs $\{x_1,..,x_m\}$ and consider the Frobenius norm $\|.\|_F$ of 2-ary functions defined only on these inputs:*

$$\|k\|_F = \sqrt{\frac{1}{m^2}\sum_{i,j=1}^m k^2(x_i,x_j)}, \text{ we have the following inequality:}$$

$$\left| \hat{A}(k+dk,h+dh,S) - \hat{A}(k,h,S) \right|$$

$$\leq 2\pi \left(\frac{\|dk\|_F}{\|k\|_F - \|dk\|_F} + \frac{\|dh\|_F}{\|h\|_F - \|dh\|_F} \right) \tag{8.72}$$

Proof. The Alignment $\hat{A}(k,h,S)$ is the cosine of the angle $\theta_{k,h}$ between k and h. Since cosine is 1-lipschitzian, variations of $\hat{A}(k,h,S)$ can be bounded by variations of θ as:

$$\left| \hat{A}(k+dk,h+dh,S) - \hat{A}(k,h,S) \right|$$

$$= \left| \cos(\theta_{k+dk,h+dh}) - \cos(\theta_{k,h}) \right| \leq \left| \theta_{k+dk,h+dh} - \theta_{k,h} \right| \tag{8.73}$$

Since the right hand side of this inequality is bounded as:

$$\left| \theta_{k+dk,h+dh} - \theta_{k,h} \right| \leq \left| \theta_{k+dk,h} \right| + \left| \theta_{h+dh,h} \right| \tag{8.74}$$

to prove bounds on variations of $\hat{A}(k,h,S)$, we are left to bound the angles $\theta_{k+dk,k}$ and $\theta_{h+dh,h}$. The latter are measured in radians. Let us consider only $\theta_{k+dk,k}$ since the second one will be bounded by following exactly the same reasoning. Consider figure 7. We would like to bound θ in this figure. We suppose here that $\|dk\|_F < \|k\|_F$. The length of the arc AC inside the circle is θOC and is smaller than the circumference $2\pi\|dk\|_F$ so that:

$$\theta \leq 2\pi \frac{\|dk\|_F}{OC} \tag{8.75}$$

But $OC \geq OB = \| k \|_F - \| dk \|_F$ so that we have finally:

$$\theta_{k+dk,k} \leq 2\pi \frac{\| dk \|_F}{\| k \|_F - \| dk \|_F} \tag{8.76}$$

Using this result in (8.74), we derive:

$$\left| \hat{A}(k+dk, h+dh) - \hat{A}(k,h) \right|$$

$$\leq 2\pi \left(\frac{\| dk \|_F}{\| k \|_F - \| dk \|_F} + \frac{\| dh \|_F}{\| h \|_F - \| dh \|_F} \right) \tag{8.77}$$

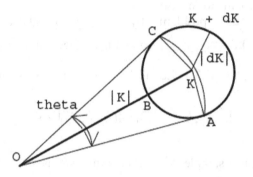

Figure 7 The angle θ denoted theta on the figure can be computed from the length of the edge OC and of the arc AC. This arc has a length smaller than $2\pi \| dK \|$. $\| dK \|$ is the radius of the circle inside which lies $K + dK$.

Theorem 19 *Let K a set of kernels, $a \in \mathbb{R}$ and h be a fixed kernel. Assume that for all $k \in K$, for any $S = \{x_1, .., x_m\}$ in χ^m, $\| k \|_S = \frac{1}{m^2} \sum_{i,j=1}^m k^2(x_i, x_j) \geq a$ and that $\| h \|_S \geq \| a \|$, then we have for $\epsilon \geq \frac{1}{a^2} \sqrt{\frac{4c_1}{m}}$*

$$\mathbb{P}\left[\sup_{k \in K} \left| \hat{A}(h,k,S) - \mathbb{E}[\hat{A}(h,k,S)] \right| \geq \epsilon \right]$$

$$\leq \mathbb{E}\left[N\left(a\epsilon/(18\pi), K, d_{S,S'} \right) \right] \exp\left(-\frac{ma^4 \epsilon^2}{c_3} \right) \tag{8.78}$$

where S' is a second set $\{x_{m+1}, .., x_{2m}\}$ sampled identically to S and $d_{S,S'}(h,k)$ is defined as:

$$d_{S,S'}(h,k) = \max_{\sigma} \sqrt{\frac{1}{m^2} \sum_{i,j=1}^{m} (h(x_{\sigma(i)}, x_{\sigma(j)}) - k(x_{\sigma(i)}, x_{\sigma(j)}))^2 +}$$

$$\sqrt{\frac{1}{m^2} \sum_{i,j=m}^{2m} (h(x_{\sigma(i)}, x_{\sigma(j)}) - k(x_{\sigma(i)}, x_{\sigma(j)}))^2} \qquad (8.79)$$

where σ is any permutation between $\{1,..,m\}$ and $\{m+1,..,2m\}$ swapping the x_i with the x_{m+i}'s: $\sigma(i) \in \{i, m+i\}$.

Proof. The reasoning we follow parallels the reasoning used for the empirical risk minimization case.

1. Symmetrize the problem by adding a ghost training set $S' = \{(x_{m+1}, y_{m+1}),..,(x_{2m}, y_{2m})\}$ so that the expectation $\mathbb{E}\left[\hat{A}(k, y, S)\right]$ is roughly equal to $\hat{A}(k, y, S')$. Let us formalize this: fix S and consider k^* such that:

$$\left|\hat{A}(k^*, h, S) - \mathbb{E}_S[\hat{A}(k^*, h, S)]\right| \geq \epsilon$$

Then using the ghost sample S' and theorem 9 we have:

$$P_{S'}\left[A(k^*, h, S') - \mathbb{E}_S[\hat{A}(k^*, h, S)] > \epsilon\right] \leq \exp(-ma^4 \epsilon^2 / c_1)$$

So that for $\epsilon \geq \frac{1}{a^2}\sqrt{\frac{4c_1}{m}}$, we have:

$$P\left[\hat{A}(k^*, h, S') - \mathbb{E}_S[\hat{A}(k^*, h, S)] \leq \frac{\epsilon}{2}\right] \geq \frac{1}{2} \qquad (8.80)$$

and finally,

$$P_{S,S'}\left[\begin{array}{l} \sup_{k \in \mathcal{K}} \left|\hat{A}(h, k, S) - \mathbb{E}_S[\hat{A}(h, k, S)]\right| \\ > \epsilon \text{ and } \hat{A}(k^*, h, S') - \mathbb{E}_S[\hat{A}(k^*, h, S)] \leq \frac{\epsilon}{2} \end{array}\right] \qquad (8.81)$$

$$=$$

$$P_{S,S'}\left[\sup_{k \in \mathcal{K}} \left|\hat{A}(h, k, S) - \mathbb{E}_S[\hat{A}(h, k, S)]\right| > \epsilon \Big| S\right]$$

$$P_{S,S'}\left[\hat{A}(k^*, h, S') - \mathbb{E}_S[\hat{A}(k^*, h, S)] > \frac{\epsilon}{2} \Big| S\right] \mathbb{P}[S]$$

But:

$$P_{S,S'}\left[\begin{array}{c} \sup_{k\in\mathcal{K}}\left|\hat{A}(h,k,S)-\mathbb{E}_S[\hat{A}(h,k,S)]\right| \\ > \epsilon \text{ and } \hat{A}(k^*,h,S')-\mathbb{E}_S[\hat{A}(k^*,h,S)]\leq\frac{\epsilon}{2} \end{array}\right] \tag{8.82}$$

$$\leq P_{S,S'}\left[\sup_{k\in\mathcal{K}}\left|\hat{A}(h,k,S)-\hat{A}(h,k,S')\right|>\frac{\epsilon}{2}\right]$$

Combined with equation (8.80), it yields:

$$P_{S,S'}\left[\sup_{k\in\mathcal{K}}\left|\hat{A}(h,k,S)-E_S[\hat{A}(h,k,S)]\right|>\epsilon\right]\leq$$

$$2P_{S,S'}\left[\sup_{k\in\mathcal{K}}\left|\hat{A}(h,k,S)-\hat{A}(h,k,S')\right|>\frac{\epsilon}{2}\right] \tag{8.83}$$

2.   Note that the problem is symmetric and that permuting (x_{m+j},y_{m+j}) with (x_j,y_j) does not change the probabilities. We can therefore add new random objects, namely random swapping permutations σ defined by: $\sigma(i)\in\{i,m+i\}$, $i=1,..,m$ and the probability that $\sigma(i)=i$ is $1/2$. We have then:

$$P_{S,S'}\left[\sup_{k\in\mathcal{K}}\left|\hat{A}(h,k,S)-\hat{A}(h,k,S')\right|\geq\frac{\epsilon}{2}\right]=$$

$$P_{S,S',\sigma}\left[\sup_{k\in\mathcal{K}}\left|\hat{A}(h,k,S_\sigma)-\hat{A}(h,k,S'_\sigma)\right|\geq\frac{\epsilon}{2}\right] \tag{8.84}$$

where $S_\sigma=\{x_{\sigma(1)},..,x_{\sigma(m)}\}$ and $S'_\sigma=\{x'_{\sigma(1)},..,x'_{\sigma(m)}\}$.

3.   We condition on the training and the ghost training set and look at the probabilities only with respect to σ:

$$P_{S,S',\sigma}\left[\sup_{k\in\mathcal{K}}\left|A(h,k,S_\sigma)-A(h,k,S'_\sigma)\right|\geq\frac{\epsilon}{2}\middle|S,S'\right]$$

We consider then a cover $\mathcal{N}$ of $\mathcal{K}$ with respect to the distance:

$$d_{S,S'}(h,k)=\max_\sigma\sqrt{\frac{1}{m^2}\sum_{i,j=1}^m(h(x_{\sigma(i)},x_{\sigma(j)})-k(x_{\sigma(i)},x_{\sigma(j)}))^2}+$$

$$\sqrt{\frac{1}{m^2}\sum_{i,j=m}^{2m}(h(x_{\sigma(i)},x_{\sigma(j)})-k(x_{\sigma(i)},x_{\sigma(j)}))^2} \tag{8.85}$$

This distance is such that: $d_{S,S'}(h,k) < \alpha \Rightarrow \|H - K\|_F < \alpha$ for any kernel matrix H and K built from h and k from the sets S_σ or S'_σ for any swapping permutation σ.

This cover is at scale $\eta = a\epsilon/(18\pi)$ so that if:

$$\sup_{k \in \mathcal{K}} \left| A(h,k,S_\sigma) - A(h,k,S'_\sigma) \right| > \frac{\epsilon}{2}$$

then, there exists k^* in $\mathcal{K}$ so that $\left| A(h,k,S_\sigma) - A(h,k,S'_\sigma) \right| \geq \frac{\epsilon}{2}$ and there exist dk such that $d_{S,S'}(dk,0) < \eta$ and $k^* = k + dk$ with $k \in \mathcal{N}$, so that: $\left| A(h,k+dk,S_\sigma) - A(h,k+dk,S'_\sigma) \right| > \frac{\epsilon}{2}$.

We use now the following inequality of Proposition 18:

$$\left| \hat{A}(k+dk,h,S_\sigma) - \hat{A}(k,h,S_\sigma) \right| \leq 2\pi \frac{\|dk\|_F}{\|k\|_F - \|dk\|_F} \qquad (8.86)$$

where $\|k\|_F = \sqrt{\frac{1}{m^2} \sum_{i,j=1}^{m} k^2(x_{\sigma(i)}, x_{\sigma(j)})}$. According to the properties of $d_{S,S'}$, we have: $\|dk\|_F \leq \eta$. We deduce:

$$\left| \hat{A}(k+dk,h,S_\sigma) - \hat{A}(k,h,S_\sigma) \right| \leq 2\pi \frac{\epsilon}{18\pi - \epsilon} \leq \frac{\epsilon}{8} \qquad (8.85)$$

So that: $\left| \hat{A}(h,k+dk,S_\sigma) - \hat{A}(h,k+dk,S'_\sigma) \right| > \frac{\epsilon}{2}$ implies $\left| \hat{A}(h,k,S_\sigma) - \hat{A}(h,k,S'_\sigma) \right| > \frac{\epsilon}{4}$. And hence:

$$P_{S,S',\sigma} \left(\sup_{k \in \mathcal{K}} \left| \hat{A}(h,k,S_\sigma) - \hat{A}(h,k,S'_\sigma) \right| \geq \frac{\epsilon}{2} \Big| S,S' \right) \leq$$
$$P_{S,S',\sigma} \left(\sup_{k \in \mathcal{N}} \left| \hat{A}(h,k,S_\sigma) - \hat{A}(h,k,S'_\sigma) \right| \geq \frac{\epsilon}{4} \Big| S,S' \right) \qquad (8.86)$$

Since $\mathcal{N}$ has a finite cardinality denoted by $\mathcal{N}(\eta,\mathcal{K},d_{S,S'})$, using the union bound with equation (8.78) we finally get the desired inequality.

9 The Structure of Version Space

Ralf Herbrich*, Thore Graepel*, Robert C. Williamson†

*Microsoft Research Ltd., Cambridge, U.K.

†National ICT Australia, Canberra, Australia

Abstract

We investigate the generalisation performance of consistent classifiers, i.e. classifiers that are contained in the so-called *version space*, both from a theoretical and experimental angle. In contrast to classical VC analysis—where no single classifier within version space is singled out on grounds of a generalisation error bound—the data dependent structural risk minimisation framework suggests that there exists one *particular* classifier that is to be preferred because it minimises the generalisation error bound. This is usually taken to provide a theoretical justification for learning algorithms such as the well known support vector machine. A reinterpretation of a recent PAC-Bayesian result, however, reveals that given a suitably chosen hypothesis space there exists a large fraction of classifiers with small generalisation error albeit we cannot identify them for a specific learning task. In the particular case of linear classifiers we show that classifiers found by the classical perceptron algorithm have guarantees bounded by the size of version space. These results are complemented with an empirical study for kernel classifiers on the task of handwritten digit recognition which demonstrates that even classifiers with a small margin may exhibit excellent generalisation. In order to perform this analysis we introduce the kernel Gibbs sampler—an algorithm which can be used to sample consistent kernel classifiers.

9.1 Introduction

Over the last ten years, machine learning has received a boost due to the ground-breaking results on the generalisation error of classifiers (see Vapnik [1982], Shawe-Taylor et al. [1998]). Their results build the theoretical basis for the well-known support vector machine (SVM) algorithm. It is now widely accepted that for complex models it is necessary to use regularisation techniques such as margin maximisation in order to find a classifier exhibiting a small generalisation error

R. Herbrich et al.: *The Structure of Version Space*, StudFuzz **194**, 257–274 (2006)

www.springerlink.com   © Springer-Verlag Berlin Heidelberg 2006

(see Vapnik [1995, p. 157]). Since for large datasets the SVM algorithm is too time consuming many heuristics to approximate the SVM solution have been put forward (see, e.g. Platt [1999], Keerthi et al. [1999], Smola and Schölkopf [2000]). Recently, it has been demonstrated experimentally that even algorithms with no explicit regularisation perform comparably to SVMs (see Mika et al. [1999], Herbrich et al. [2001]). This observation raises an interesting question:

> What fraction of classifiers within version space exhibit a small generalisation error?

In this paper we try to answer this question both from a theoretical and experimental point of view. Using a recent result in the PAC-Bayesian framework we are able to show that given a suitably chosen hypothesis space there exists a large fraction of classifiers with small generalisation error. More precisely, *the generalisation error of most of the classifiers in version space is controlled by the size of the version space relative to the size of the hypothesis space.* This result, which we call the *egalitarian* generalisation error bound, is complemented by an experimental study for linear classifiers on the task of handwritten digit recognition using the MNIST database. It is worthwhile mentioning that in a fully Bayesian treatment the size of version space is also called the *evidence* of the model or hypothesis space, respectively (see MacKay [1992]).

The paper is structured as follows: in the following section we review generalisation error bounds for single classifiers consistent with the whole training sample. We will also introduce the PAC-Bayesian framework and its main result which allows us to give our main theoretical result together with its proof at the end of this section. In the subsequent section we discuss the impact of this result for practical learning theory. We also give a more specific result for the perceptron learning algorithm that points into the same direction. In Section 4 we present the kernel Gibbs sampler algorithm which allows us to validate our theoretical result on a benchmark problem in the field of handwritten digit recognition. The paper concludes with a discussion of generalisation error bounds for specific algorithms as opposed to bounds that hold *uniformly* over version space.

We denote a probability measure by $\mathbf{P_X}$; random variables are typeset in upper capital sans-serif font. The symbols $\mathbf{E}$ and $\mathbf{I}$ denote the expectation of a random variable and the indicator function, respectively. We use bold roman font for vectors $\mathbf{x}$ and denote tuples by $\boldsymbol{x}$. Finally, the symbol ℓ_2^n denotes the space of all sequences $\mathbf{x} = (x_1, \ldots, x_n)$ of length n for which $\sum_{i=1}^{n} x_i^2 < \infty$.

9.2 Generalisation Error Bounds for Consistent Classifiers

Suppose we are given a sample $\boldsymbol{x} = (x_1, \ldots, x_m) \in \mathcal{X}^m$ together with a sample $\boldsymbol{y} = (y_1, \ldots, y_m) \in \mathcal{Y}^m = \{-1, +1\}^m$ drawn iid from an unknown distribution $\mathbf{P_Z} = \mathbf{P_{XY}}$. Furthermore, assume we are given a fixed *hypothesis space* $\mathcal{H}$ of functions $h : \mathcal{X} \to \mathcal{Y}$. We consider learning algorithms that aim at finding a function $h^* \in \mathcal{H}$ that minimises the *generalisation error* $R[h]$ given by

$$R[h] = \mathbf{P_{XY}}(h(\mathsf{X}) \neq \mathsf{Y}) = \mathbf{E_{XY}}\left[\mathbf{I}_{h(\mathsf{X}) \neq \mathsf{Y}}\right].$$

A common approach to (approximately) finding h^* based on the training sample $\boldsymbol{z} = (\boldsymbol{x}, \boldsymbol{y}) \in \mathcal{Z}^m$ is to select a function $h \in \mathcal{H}$ that minimises the *training error* $R_{\mathrm{emp}}[h, \boldsymbol{z}]$

$$R_{\mathrm{emp}}[h, \boldsymbol{z}] = \frac{1}{m} \sum_{(x_i, y_i) \in \boldsymbol{z}} \mathbf{I}_{h(x_i) \neq y_i}.$$

Let us assume that $\mathbf{P_{Y|X=x}}(y) = \mathbf{I}_{h^*(x) = y}$, i.e. h^* deterministically labels all the data and thus has minimal generalisation error. Then we define the *version space* $V(\boldsymbol{z})$ (phrase due to Mitchell [1982]) as the set of all classifiers $h \in \mathcal{H}$ that are *consistent* with the training sample $\boldsymbol{z}$,

$$V(\boldsymbol{z}) = \{h \in \mathcal{H} \mid R_{\mathrm{emp}}[h, \boldsymbol{z}] = 0\}.$$

Of course, solely based on the training error $R_{\mathrm{emp}}[h, \boldsymbol{z}]$ all classifiers in version space are indistinguishable. Moreover, even if a classifier has zero training error it can happen that its generalisation error is large—an effect known as *over-fitting*. In order to cope with this uncertainty a lot of research has been done to obtain probabilistic bounds on the generalisation error of consistent classifiers. The basic idea is to guarantee that for most training trials (random training samples) the generalisation error of a consistent classifier does not exceed a certain value.

Definition 1 (PAC Generalisation Error Bound). A function $\varepsilon : \mathbb{N} \times \mathcal{H} \times \cup_{m=1}^{\infty} \mathcal{Z}^m \times [0, 1] \to \mathbb{R}$ such that for all measures $\mathbf{P_Z}$, for all $m \in \mathbb{N}$ and for all $\delta \in (0, 1]$

$$\mathbf{P_{Z^m}}(\forall h \in \mathcal{H} : (h \notin V(\mathbf{Z})) \vee (R[h] < \varepsilon(m, h, \mathbf{Z}, \delta))) \geq 1 - \delta \qquad (1)$$

is called a *PAC generalisation error bound* for the hypothesis space $\mathcal{H}$.

Classical VC theory (see Vapnik [1982, 1995]) provides the following bound for all $m > d_{\mathcal{H}}$ and for all hypotheses $h \in \mathcal{H}$:

$$\varepsilon_{\mathrm{VC}}(m, h, \boldsymbol{z}, \delta) = \varepsilon_{\mathrm{VC}}(m, \delta) = \frac{4}{m}\left(\ln\left(\left(\frac{2em}{d_{\mathcal{H}}}\right)^{d_{\mathcal{H}}}\right) + \ln\left(\frac{2}{\delta}\right)\right), \qquad (2)$$

where $d_{\mathcal{H}}$ is known as the *VC dimension* of the hypothesis space $\mathcal{H}$ (see Vapnik [1982] for more details). Obviously, the generalisation error *bound* is *independent* of the particular classifier $h \in V(z)$ and as such no single classifier $h \in V(z)$ is singled out on the basis of VC theory.

However, in applied classification learning it is common practice that the classification is carried out by thresholding a real-valued function, i.e. $h(x) = \text{sign}(f(x))$. It can be shown that the additional information of the real-valued magnitude $|f(x)|$ *before* thresholding allows one to obtain a generalisation error bound in terms of the margin $\gamma_z(h) = \min_{(x_i, y_i) \in z} y_i f(x_i)$ attained on the given sample z, i.e., for all hypotheses $h \in \mathcal{H}$ and $m > d_{\mathcal{H}}(\tilde{\gamma}_z(h))$, $\tilde{\gamma}_z(h) := \gamma_z(h)/8$

$$\varepsilon_{\text{fat}}(m, h, z, \delta) = \varepsilon_{\text{fat}}(m, \tilde{\gamma}_z(h), \delta)$$

$$= \frac{2}{m}\left(\log_2\left(\left(\frac{8em}{d_{\mathcal{H}}(\tilde{\gamma}_z(h))}\right)^{d_{\mathcal{H}}(\tilde{\gamma}_z(h))}\right)\log_2(32m) + \log_2\left(\frac{2m}{\delta}\right)\right), \quad (3)$$

where $d_{\mathcal{H}}(\gamma)$ is known as the *fat shattering* dimension of the hypothesis space $\mathcal{H}$ at the observed scale γ (see Shawe-Taylor et al. [1998], Kearns and Schapire [1994] for details). The function $d_{\mathcal{H}} : \mathbb{R}^+ \to \mathbb{N}$ is always monotonically non-increasing and is a straightforward generalisation of the VC dimension to sets of real valued functions. An immediate consequence of this result is that the *bound on the* generalisation error $R[h]$ depends inversely on the margin $\gamma_z(h)$. As such the result singles out *one* classifier within version space — the classifier with maximal margin also known as the support vector solution (see Vapnik [1995]).

Recently, McAllester [1998] presented "some PAC–Bayesian theorems" which provide a generalisation error bound for the Gibbs classification strategy Gibbs_z. Given a prior $\mathbf{P}_{\mathsf{H}}$ over hypothesis space $\mathcal{H}$ and a training sample z, for each test example x the Gibbs classification strategy samples a classifier $h \in V(z)$ according to $\mathbf{P}_{\mathsf{H}|\mathsf{H} \in V(z)}$ and uses it for classification $\text{Gibbs}_z(x)$. Note that Gibbs_z does not correspond to any *single* classifier $h \in V(z)$ but to a classification strategy based on $\mathbf{P}_{\mathsf{H}|\mathsf{H} \in V(z)}$. For any prior $\mathbf{P}_{\mathsf{H}}$, the PAC bound $\varepsilon_{\text{Gibbs}}$ on the generalisation error $R[\text{Gibbs}_z] = \mathbf{E}_{\mathsf{H}|\mathsf{H} \in V(z)}[R[H]]$ of this stochastic classification strategy is given by

$$\varepsilon_{\text{Gibbs}}(m, \mathbf{P}_{\mathsf{H}}, z, \delta) = \frac{1}{m}\left(\ln\left(\frac{1}{\mathbf{P}_{\mathsf{H}}(V(z))}\right) + \ln\left(\frac{em^2}{\delta}\right)\right), \quad (4)$$

hence

$$\mathbf{P}_{Z^m}\left(R[\text{Gibbs}_Z] \leq \varepsilon_{\text{Gibbs}}(m, \mathbf{P}_{\mathsf{H}}, Z, \delta)\right) \geq 1 - \delta. \quad (5)$$

The first term in (2)—which is driven by the worst case number of equivalence classes w.r.t. the two classes $y \in \mathcal{Y}$—has been replaced by a *data-dependent* quantity—the prior belief $\mathbf{P_H}$ in consistent classifiers $h \in V(z)$. As opposed to classical PAC generalisation error bounds, this result *does not provide any guarantee for single classifiers* $h \in V(z)$. The first theoretical result of the present paper is a direct consequence of (4) and is stated in the following theorem.

Theorem 2 (Egalitarian Bound). *For all measures* $\mathbf{P_Z}$, *with probability at least* $1 - \delta$ *over the random draw of the training sample* z *of size* m, *for all* $\eta > 1$, *at least a fraction of* $1 - \frac{1}{\eta}$ *of the classifiers in version space* $V(z)$ *have generalisation error less than*

$$\eta \cdot \varepsilon_{\text{Gibbs}}(m, \mathbf{U_H}, z, \delta) \, ,$$

where $\mathbf{U_H}$ *is the uniform measure over* $\mathcal{H}$.

Proof. The proof is a simple application of Markov's inequality along with the instantiation of $\mathbf{P_H}$ by the uniform measure $\mathbf{U}_{\mathcal{H}}$. Markov's inequality says

$$\forall \eta > 1: \qquad \mathbf{P}_{\mathsf{H}|\mathsf{H} \in V(\mathbf{Z})}\left(R\left[\mathsf{H}\right] \geq \eta \cdot \mathbf{E}_{\mathsf{H}|\mathsf{H} \in V(\mathbf{Z})}\left[R\left[\mathsf{H}\right]\right]\right) < \frac{1}{\eta},$$

because the generalisation error $R : \mathcal{H} \to [0, 1]$ as a functional over hypotheses is a positive random variable. Hence, from (5) it follows

$$\mathbf{P}_{\mathbf{Z}^m}\left(\forall \eta > 1 : \mathbf{P}_{\mathsf{H}|\mathsf{H} \in V(\mathbf{Z})}\left(R\left[\mathsf{H}\right] < \eta \cdot \varepsilon_{\text{Gibbs}}(m, \mathbf{U_H}, \mathbf{Z}, \delta)\right) \geq 1 - \frac{1}{\eta}\right) \geq 1 - \delta.$$

$\square$

In the following section we shall discuss this results and its impact on the structure of version space. However, one of the most intriguing features of this generalisation error bound is that it holds true regardless of any property of the single classifiers considered. In fact, the only quantity that drives the generalisation error bound is the volume of version space which is a *property of the model* $\mathcal{H}$ *and the data* z but not of single classifiers h.

9.3 Consequences of the Egalitarian Bound

9.3.1 Linear Classifiers

Consider the result of Theorem 2 with $\eta = 2$ and the hypothesis space $\mathcal{H}$ used in SVMs. In this case we know that with high probability ($\geq 1-\delta$) the generalisation error of at least half of the classifiers in version space $V(z)$ are bounded by at most twice the generalisation error of the Gibbs classification strategy. This should be compared with a typical generalisation error bound for linear classifiers in terms of margins (see Herbrich and Graepel [2002])

$$\frac{2}{m}\left(\ln\left(\frac{2}{\Gamma_{\mathbf{z}}^2(h)}\right)^n + \ln\left(\frac{(em)^2}{\delta}\right)\right) \geq 2 \cdot \varepsilon_{\text{Gibbs}}\left(m, \mathbf{U_H}, \mathbf{z}, \delta\right). \tag{6}$$

Here, n is the dimensionality of the feature space $\mathcal{K} \subseteq \ell_2^n$ in which the linear classification is carried out. The first term is the inverse of a lower bound on the volume of version space $V(\mathbf{z})$ in terms of a *normalised margin* $\Gamma_{\mathbf{z}}(h)$ given by

$$\Gamma_{\mathbf{z}}(h) \propto \min_{(x_i, y_i) \in \mathbf{z}} \frac{y_i f(x_i)}{\|x_i\|}, \tag{7}$$

which coincides with $\gamma_{\mathbf{z}}(h)$ for normalised data only. Thus we see that *whenever the SVM solution has a small generalisation error bound at least half of the consistent classifiers have the same (or even better) generalisation error bound.* The practical difficulty in exploiting these solutions, however, is that they keep changing over the random draw of the training sample and only the large margin classifier is able to *witness* its small generalisation error by an easy-to-determine quantity—its margin. Nonetheless, randomly drawing a consistent classifier will do as well in at least half of the learning trials *if the hypothesis space (model)* was suited for the task at hand. The result suggests one should not be too dismissive of algorithms such as the perceptron learning algorithm [Rosenblatt, 1958] which merely ensure one gets an $h \in V(\mathbf{z})$. It appears that the choice of the model $\mathcal{H}$ is more important than the choice of the learning procedure *within a fixed model $\mathcal{H}$.* For kernel based classifiers this means the choice of the kernel (see also Section 4).

9.3.2   From Margin To Sparsity—A Revival of the Perceptron

Theorem 2 tells us that whenever the training sample $\mathbf{z}$ observed and the hypothesis space $\mathcal{H}$ chosen lead to a large version space, there *exists* a large fraction of classifiers $h \in V(\mathbf{z})$ with a small generalisation error. In the special case of linear classifiers there is also an efficient algorithm for finding some of these classifiers — the perceptron algorithm [Rosenblatt, 1958]. In particular, we can prove the following theorem (see Graepel et al. [2001], Gat [2001] for more details).

Theorem 3 (Margin Bound). *For any measure $\mathbf{P_Z}$, with probability at least $1 - \delta$ over the random draw of the training set $\mathbf{z} = (\mathbf{x}, \mathbf{y}) \in (\mathcal{X} \times \{-1, +1\})^m$ of size m, if there exists a linear classifier $h^* \in \mathcal{H}$ such that*

$$\kappa^* = \left\lceil \frac{1}{\Gamma_{\mathbf{z}}^2(h^*)} \right\rceil \leq m$$

*then the generalisation error $R[h]$ of the classifier $h \in V(z)$ found by the percep-
tron algorithm is less than*

$$\frac{1}{m - \kappa^*} \left(\ln \left(\binom{m}{\kappa^*} \right) + \ln(m) + \ln \left(\frac{1}{\delta} \right) \right). \tag{8}$$

Proof. The proof is a combination of a results in Novikoff [1962] on the number of mistakes of the perceptron learning algorithm and a compression bound
(see Littlestone and Warmuth [1986], Floyd and Warmuth [1995], Graepel et al.
[2000b]). At first, Novikoff's theorem tells us that for normalised data $x \in \mathcal{X}^m$
the perceptron learning algorithm is guaranteed to make at most κ^* mistakes.
At each mistake, it adds (or subtracts) the current data point x_i to the weight
vector which was initially set to $\mathbf{0}$. As a consequence thereof, the number of
training samples (x_i, y_i) used to construct the final hypothesis is always less than
or equal to κ^*. Since there are at most $\binom{m}{\kappa^*}$ different subsets of training samples
of size κ^* the effective number of different hypotheses $h \in V(z)$ is this number.
A combination of the binomial tail bound on the $m - \kappa^*$ left-out training points,
i.e.

$$\forall h \in \mathcal{H}: \quad \mathbf{P}_{\mathbf{Z}^{m-\kappa^*}} \left((h \notin V(\mathbf{Z})) \vee \left(R[h] \leq \frac{\ln \left(\frac{1}{\delta} \right)}{m - \kappa^*} \right) \right) \geq 1 - \delta,$$

with the union bound over the number of different subsets proves the theorem.
Note that the additional $\ln(m)$ term is due to the fact that the value of κ^* is not
fixed. This requires us to share the confidence of $1 - \delta$ among all its at most m
different values. $\square$

Similar to the egalitarian bound this result is somewhat surprising as the generalisation error of the classifier learned by the perceptron learning algorithm is
controlled by the potential margin $\Gamma_z(h^*)$ a SVM *would have achieved* on the
same training sample z. Combining this result with the fact that margin bounds
for support vector machines just witness the good choice of a model $\mathcal{H}$ (see (6))
we conclude that the simple perceptron algorithm is theoretically well justified
because *whenever the SVM solution has a small generalisation error bound all
the up to m! different classifiers learned with the perceptron learning algorithm
have the same (or even better) generalisation error bound.* This has also found
some empirical evidence in the binary classification problems of handwritten digit
recognition (see Freund [1999]).

9.3.3 Bayes Classification Strategy

Another consequence of Theorem 2 is that half of the classifiers within version
space $V(z)$ have a generalisation error bound as good as that of the Bayes classification strategy. The Bayes classification strategy—also known as Bayesian

transduction (see Vapnik [1982], Graepel et al. [2000a])—assigns a test example x to the class y by majority voting under the measure $\mathbf{P}_{\mathsf{H}|\mathsf{H}\in V(\boldsymbol{z})}$,

$$Bayes_{\boldsymbol{z}}(x) = \operatorname{argmax}_{y\in\mathcal{Y}} \mathbf{P}_{\mathsf{H}|\mathsf{H}\in V(\boldsymbol{z})}(\mathsf{H}(x) = y) \ .$$

In contrast to the Gibbs classification strategy, the Bayes classification strategy *deterministically* assigns a new test example to a class. For $|\mathcal{Y}| = 2$, whenever the Bayes classification strategy is wrong at x, at least half of the classifiers in version space misclassify x, too. By this argument, the generalisation error bound of the Bayes classification strategy fulfils

$$\forall \mathbf{P}_{\mathsf{H}} : \qquad \varepsilon_{\text{Bayes}}(m, \mathbf{P}_{\mathsf{H}}, \boldsymbol{z}, \delta) \leq 2 \cdot \varepsilon_{\text{Gibbs}}(m, \mathbf{P}_{\mathsf{H}}, \boldsymbol{z}, \delta) \ . \qquad (9)$$

This equivalence of generalisation error bounds finds empirical support in Graepel et al. [2000a], Herbrich et al. [2001]. Note that the "averaging" and "voting" feature of the Gibbs and Bayes strategies, respectively, safeguards them against domination by a minority of inferior members of the version space $V(\boldsymbol{z})$.

9.3.4   Have we Thrown the Baby out with the Bath Water?

At first glance the egalitarian bound seems to imply that we are hopeless in the search for *the* quantity controlling generalisation error (bounds) because it gives a good generalisation error bound for a huge number of consistent classifiers $h \in V(\boldsymbol{z})$ not referring to any property other than the choice of the model $\mathcal{H}$. This result, however, comes at no surprise taking into account what we investigated theoretically (see Definition 1). Although one is typically only interested in the performance of the one classifier h learned using a fixed learning algorithm $\mathcal{A} : \cup_{m=1}^{\infty} \mathcal{Z}^m \to \mathcal{H}$ traditional learning theory claims to need guarantees on the generalisation error that hold *uniformly* over the whole hypothesis space $\mathcal{H}$ or version space $V(\boldsymbol{z})$, respectively. This is much too demanding and can therefore only lead to bounds that indicate whether we have chosen an appropriate model or not. A much more promising approach seems to investigate the question of generalisation error bounds for specific algorithms. In fact, the proof of Theorem 3 uses a compression bound which requires the specification of the algorithm $\mathcal{A}$ in advance, i.e., the bounds apply only to a small subset of learning algorithms (so called *compression schemes*). A related idea is studied in Bousquet and Elisseeff [2001] where the VC dimension as a complexity measure of an hypothesis space $\mathcal{H}$ is replaced by the *robustness* of the learning algorithm $\mathcal{A}$ used. The robustness of an algorithm $\mathcal{A}$ measures by how much the training error of the learned classifier $\mathcal{A}(\boldsymbol{z})$ is changing when adding one additional observation, i.e. $\max_{z=(x,y)} |R_{\text{emp}}[\mathcal{A}(\boldsymbol{z}), \boldsymbol{z}] - R_{\text{emp}}[\mathcal{A}(\boldsymbol{z} \cup z), \boldsymbol{z} \cup z]|$. According to intuition, whenever a learning algorithm is very robust we have small deviation between generalisation and training error for the classifiers learned although the VC dimension of the hypothesis class used might have been infinite.

Finally, it is worthwhile noticing that this result does not deny the importance of *inductive principles*. Although we know that within a good model $\mathcal{H}$ there are many classifiers with a provably small generalisation error, there might exist procedures (the maximum margin algorithm is one such procedure) that single out classifiers with small generalisation error bounds for most random draws of the training sample $\mathbf{z}$. A potential candidate for formulating such inductive principles is the *luckiness framework* Shawe-Taylor et al. [1998], which was recently extended to include an explicit dependency on the learning algorithm Herbrich and Williamson [2002].

9.4 Experimental Results for Linear Classifiers

In order to complement the above theoretical analysis let us empirically evaluate the distribution of generalisation errors over version space members. Let us consider the hypothesis class $\mathcal{H}$ provided by linear classifiers in feature space $\mathcal{K} \subseteq \ell_2^n$ as used in SVMs. Each hypothesis is given by

$$h_{\mathbf{w}}(x) = \mathrm{sign}\left(\langle \phi(x), \mathbf{w}\rangle\right) = \mathrm{sign}\left(\langle \mathbf{x}, \mathbf{w}\rangle\right), \tag{10}$$

where $\phi : \mathcal{X} \to \mathcal{K} \subseteq \ell_2^n$ is a mapping[1] from the input space $\mathcal{X}$ to the feature space $\mathcal{K}$. Note that it is sufficient to consider weight vectors $\mathbf{w} \in \mathcal{K}$ of unit length, i.e. $\mathbf{w} \in \mathcal{W}$, $\mathcal{W} = \{\mathbf{w} \in \mathcal{K} \mid \|\mathbf{w}\| = 1\}$, because for any positive constant

$$\forall \lambda > 0 : \qquad h_{\mathbf{w}} = \mathrm{sign}\left(\langle \mathbf{x}, \mathbf{w}\rangle\right) = \mathrm{sign}\left(\langle \mathbf{x}, \lambda\mathbf{w}\rangle\right) = h_{\lambda\mathbf{w}}.$$

Ergo, the hypothesis space $\mathcal{H}$ is isomorphic to the unit sphere $\mathcal{W} \subset \ell_2^n$ (see also Figure 1). If the objective function optimised by the learning algorithm depends only on the inner products of the weight vector $\mathbf{w}$ with all the mapped training points it can be shown that it is sufficient to consider normal vectors $\mathbf{w} \in \mathcal{W}$ that are linearly expandable in the training points [Kimeldorf and Wahba, 1970, Schölkopf et al., 2001],

$$\mathbf{w} = \sum_{i=1}^{m} \alpha_i \mathbf{x}_i .$$

As a consequence, each hypothesis h can be written in terms of $\boldsymbol{\alpha} \in \mathbb{R}^m$, i.e.

$$h_{\boldsymbol{\alpha}}(x) = \mathrm{sign}\left(\sum_{i=1}^{m} \alpha_i \langle \mathbf{x}_i, \mathbf{x}\rangle\right) = \mathrm{sign}\left(\sum_{i=1}^{m} \alpha_i k(x_i, x)\right),$$

---

[1] We abbreviate $\phi(x)$ by $\mathbf{x}$ always assuming ϕ to be fixed. This, however, should not be confused with the training sample $\boldsymbol{x} \in \mathcal{X}^m$.

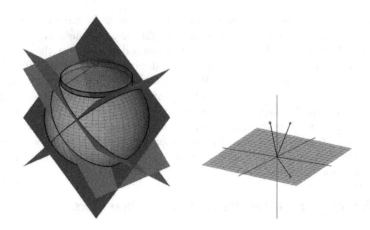

Fig. 1: **(Left)** The hypothesis space $\mathcal{H}$ of linear classifiers for a 3–dimensional feature space $\mathcal{K}$. Each point on the unit sphere is the weight vector $\mathbf{w} \in \mathcal{W}$ of a linear classifier $h_{\mathbf{w}}$ (see (10)). The convex polyhedron on top is a version space $V(\mathbf{z})$; the length of the gray line is proportional to the normalised margin $\Gamma_{\mathbf{z}}(h_{\mathbf{w}})$ of the classifier on top of the sphere. **(Right)** Three data points $\mathbf{x}_1$, $\mathbf{x}_2$ and $\mathbf{x}_3$ in a 3–dimensional feature space $\mathcal{K} \subseteq \ell_2^3$. Note that the planes in the left picture are incurred by each of the three training points by $\{\mathbf{w} \in \mathcal{K} \mid \langle \mathbf{x}, \mathbf{w} \rangle = 0\}$. Using exactly the same rule, each point $\mathbf{w} \in \mathcal{W}$ on the unit sphere in the left picture induces a decision plane $\{\mathbf{x} \in \mathcal{K} \mid \langle \mathbf{x}, \mathbf{w} \rangle = 0\}$ in feature space.

where the inner product function $k : \mathcal{X} \times \mathcal{X} \rightarrow \mathbb{R}$ is also known as the *kernel* (see, e.g. Vapnik [1995]). In practical application, it is often more convenient to select the kernel than the feature mapping ϕ.

9.4.1 The Kernel Gibbs Sampler

In order to sample consistent classifiers uniformly from $V(\mathbf{z})$ we suggest a Markov Chain sampling method known as the *kernel Gibbs*[2] *sampler* [Graepel and Herbrich, 2001]. It is a variant of the well-known hit-and-run sampling algorithm Smith [1984], which was recently shown to exhibit a fast mixing time of $\mathcal{O}(n^3)$, where n is the dimensionality of the space Lovasz [1999]. The kernel Gibbs sampler is applicable whenever $\mathbf{P}_{\mathsf{H}|\mathsf{Z}^m = \mathbf{z}}$ is a piecewise constant density proportional

---

[2] This should not be confused with the *Gibbs classification strategy*.

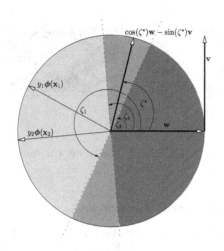

Two data points $y_1\mathbf{x}_1$ and $y_2\mathbf{x}_2$ divide the space of normalised weight vectors $\mathbf{w} \in \mathcal{W}$ into four equivalence classes with different posterior density indicated by the gray shading. In each iteration, starting from $\mathbf{w}_{j-1}$ a random direction $\mathbf{v}$ with $\mathbf{v}\perp\mathbf{w}_{j-1}$ is generated. We sample from the piecewise constant density on the great circle determined by the plane defined by $\mathbf{w}_{j-1}$ and $\mathbf{v}$. In order to obtain ζ^*, we calculate the $2m$ angles ζ_i where the training samples intersect with the circle and keep track of the number $m \cdot e_i$ of training errors for each region i.

Fig. 2: Schematic view of the kernel Gibbs sampling procedure.

to

$$\mathcal{L}[h, \mathbf{z}] = \theta^{m \cdot R_{\mathrm{emp}}[h, \mathbf{z}]} (1 - \theta)^{m(1 - R_{\mathrm{emp}}[h, \mathbf{z}])} , \quad \text{for some } \theta \in [0, 1] . \tag{11}$$

Note that this density arises from a Bayesian consideration of learning when assuming that the classification is corrupted by label noise of level $\theta \in [0, 1]$, i.e.

$$\mathbf{P}_{\mathsf{Y}|\mathsf{X}=x,\mathsf{H}=h} (y) = \theta \cdot \mathbf{I}_{y \neq h(x)} + (1 - \theta) \mathbf{I}_{y = h(x)} . \tag{12}$$

For a given value of the noise level θ and an arbitrary starting point $\mathbf{w}_0 \in \mathcal{W}$, the sampling scheme can be decomposed into the following steps (see also Figure 2):

1. Choose a direction $\mathbf{v} \in \mathcal{W}$ in the tangent space $\{\tilde{\mathbf{v}} \in \mathcal{W} \mid \langle \tilde{\mathbf{v}}, \mathbf{w}_j \rangle = 0\}$.

2. Calculate all m hit points $\mathbf{b}_i \in \mathcal{W}$ from $\mathbf{w}$ in direction $\mathbf{v}$ with the hyperplane having normal $y_i\mathbf{x}_i$. Before normalisation, this is achieved by [Herbrich et al., 2001]

$$\mathbf{b}_i = \mathbf{w}_j - \frac{\langle \mathbf{w}_j, \mathbf{x}_i \rangle}{\langle \mathbf{v}, \mathbf{x}_i \rangle} \mathbf{v} .$$

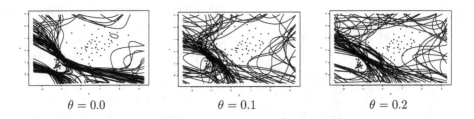

$\theta = 0.0$ $\theta = 0.1$ $\theta = 0.2$

Fig. 3: A set of 50 samples $\mathbf{w}_j$ for various noise levels θ. Shown are the resulting decision boundaries in input space $\mathcal{X} = \mathbb{R}^2$.

3. Calculate the $2m$ angular distances ζ_i from the current position $\mathbf{w}_j$.

4. Sort the ζ_i in ascending order (resulting in a permutation $\Pi : \{1, \ldots, 2m\} \to \{1, \ldots, 2m\}$ and calculate the training errors $e_i = R_{\mathrm{emp}}[h_{\mathbf{m}_i}, \mathbf{z}]$ of the $2m$ intervals $[\zeta_{\Pi(i-1)}, \zeta_{\Pi(i)}]$ by evaluating

$$\mathbf{m}_i = \cos\left(\frac{\zeta_{\Pi(i+1)} - \zeta_{\Pi(i)}}{2}\right)\mathbf{w}_j - \sin\left(\frac{\zeta_{\Pi(i+1)} - \zeta_{\Pi(i)}}{2}\right)\mathbf{v}.$$

Here, we have defined $\zeta_{\Pi(2m+1)} = \zeta_{\Pi(1)}$.

5. Sample an angle ζ^* using the piecewise uniform distribution and (11). Calculate a new sample $\mathbf{w}_{j+1}$ by $\mathbf{w}_{j+1} = \cos(\zeta^*)\mathbf{w}_j - \sin(\zeta^*)\mathbf{v}$.

6. Set $j \leftarrow j + 1$ and go back to step 1.

Since the algorithm is carried out in feature space $\mathcal{K}$ we use

$$\mathbf{w} = \sum_{i=1}^{m} \alpha_i \mathbf{x}_i, \quad \mathbf{v} = \sum_{i=1}^{m} \nu_i \mathbf{x}_i, \quad \mathbf{b} = \sum_{i=1}^{m} \beta_i \mathbf{x}_i.$$

For the inner products and norms it follows that $\langle \mathbf{w}, \mathbf{v} \rangle = \boldsymbol{\alpha}' \mathbf{G} \boldsymbol{\nu}$, $\|\mathbf{w}\|^2 = \boldsymbol{\alpha}' \mathbf{G} \boldsymbol{\alpha}$, where the $m \times m$ matrix $\mathbf{G}$ is known as the *kernel* or *Gram matrix* and is given by $\mathbf{G}_{ij} = \langle \mathbf{x}_i, \mathbf{x}_j \rangle = k(x_i, x_j)$. In Figure 3 we have shown an application of the kernel Gibbs sampler to some toy data in $\mathbb{R}^2$. As can be seen from these plots, increasing the noise level θ leads to more diverse classifiers on the training sample $\mathbf{z}$. In the following we will fix the noise level θ to zero in order to sample version space classifiers only. Other applications of this sampling algorithm are active learning, transduction and confidence estimation with kernel classifiers.

9.4.2   Distribution of Generalisation Errors and Margins

Based on the MNIST dataset[3] for images of "1" and "2" we generated well-balanced training and test samples of size 118 and 453, respectively. In order to explore the structure of version space we were interested in the distribution of generalisation errors (estimated on the given test sample) *and* its relation to the attained margin $\Gamma_z (h)$. In Figure 4 (left) we plotted the distribution of generalisation errors for $l = 10000$ samples $\mathbf{w}$ using different degrees of the polynomial kernel

$$k (x_i, x_j) = \left(\langle x_i, x_j \rangle_{\mathcal{X}} + 1 \right)^p , \tag{13}$$

which produced excellent classifiers when used in SVM learning ($p_{\mathrm{opt}} = 5$). In order to reduce dependencies between successive samples $\mathbf{w}$ of the Markov chain we used only one in ten samples thus effectively having $l_{\mathrm{eff}} = 1000$ samples. For any value of p considered there are at least 50% of consistent classifiers whose generalisation error is smaller than the one found by the SVM ($\triangle$) in accordance with (6) and the egalitarian bound of Theorem 2. Surprisingly, with increasing polynomial degree p the variance of the distribution keeps decreasing while only a small increase of its mean can be observed beyond degree 5. Furthermore, using the Bayes point machine algorithm that returns the "centre of mass" of version space $V (z)$ (see Herbrich et al. [2001]) or the SVM on the normalised training sample in feature space $\mathcal{K}$ we seem to be able to find classifiers always within the best 50% ($\circ$ and $\times$). Both these algorithms aim at finding a solution at the "centre" of version space $V (z)$ in the sense of Γ_z (see (7)).

In Figure 4 (right) we additionally provide the distributions of generalisation error for given attained margins $\Gamma_z (h)$. As expected, *almost all of the classifiers h with a large margin $\Gamma_z (h)$ do have a small generalisation error $R [h]$*. The plot also clarifies that large margins are only (probabilistically) a *sufficient condition* for good generalisation ability and that there exist many consistent classifiers with good generalisation error despite of their small margins. This is again in accordance with the egalitarian bound of Theorem 2 keeping in mind that in high-dimensional feature spaces $\mathcal{K}$ the uniform measure over volumes is concentrated near the edges. Hence, most of the classifiers in version space $V (z)$ *do have* a small margin (see width of the box-plots) albeit exhibiting good generalisation.

9.5   Conclusion

The notion of version space plays a crucial rule both in the theoretical analysis of learning algorithms and in their practical implementation. We have presented

---

[3] publicly available at `http://www.research.att.com/~yann/ocr/mnist/`.

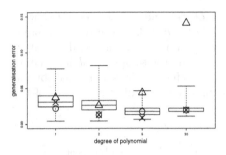 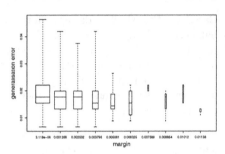

Fig. 4: **(Left)** Box-plots of distributions of generalisation errors for $l = 1000$ samples **w** using different degrees in the polynomial kernel (13). The $\triangle$, $\times$ and $\circ$ depict the generalisation errors of the SVM solution, the SVM solution when normalising in feature space $\mathcal{K}$ and the Bayes point machine solution (see text), respectively. **(Right)** Box-plots of distributions of generalisation for different attained margins (7) when using a polynomial kernel of degree 5. The width of each box-plot is proportional to the number of samples on which it is based.

a theorem which shows that within a wisely chosen hypothesis space many consistent classifiers show good generalisation irrespective of the maximisation of a pre-specified complexity measure (luckiness) such as margin. Our empirical results strongly support this conclusion and give an intuition for the structure of version space.

While the restriction to zero training-error classifiers may appear to be severe at first glance, for linear classifiers this limitation is easily overcome by modifying the kernel as follows:

$$k_\lambda(x_i, x_j) = k(x_i, x_j) + \lambda \mathbf{I}_{x_i = x_j}.$$

This trick—well known in SVMs as the quadratic soft-margin technique [Cortes and Vapnik, 1995]—gradually (with increasing λ) decouples the training examples $\phi(x_i)$ for learning and thus serves to create a version space even if the training examples were not separable under the original kernel k. Furthermore, it is straightforward to exploit Theorem 2 of McAllester [1998] so as to generalise the egalitarian bound to *any* subset H of hypothesis space $\mathcal{H}$. The difference to the present result is that in this case for many classifiers the generalisation error is effectively bounded by the training error plus the penalty $-\ln(\mathbf{P}_\mathsf{H}(H))$.

In case most of the classifiers in hypothesis space exhibit a small training error $(\mathbf{P}_H (H) \approx 1)$ we see that we get a conceptually similar result to Theorem 2. Hence, our results also cover certain cases of inconsistent classifiers deemed so important in practice.

It is worthwhile mentioning that a consequence of the above mentioned generalisation of Theorem 2 is that with high probability over the random draw of the training sample for many classifiers in *hypothesis space* the deviation between generalisation and training error is small. This result holds regardless of the VC dimension of hypothesis space $\mathcal{H}$ used. The challenge is to find generalisation error bounds that indicate if this result also holds for the *single* classifier we learned from the observed training sample.

Acknowledgements

We would like to thank Olivier Bousquet and Matthias Seeger for careful proof-reading and many useful suggestions. Furthermore, we are greatly indebted to John Shawe-Taylor, Peter Bartlett, Jonathan Baxter and Martin Anthony for helpful discussions and comments. This work was supported by the Australian Research Council.

References

O. Bousquet and A. Elisseeff. Algorithmic stability and generalization performance. In T. K. Leen, T. G. Dietterich, and V. Tresp, editors, *Advances in Neural Information Processing Systems 13*, pages 196–202. MIT Press, 2001.

C. Cortes and V. Vapnik. Support vector networks. *Machine Learning*, 20:273–297, 1995.

S. Floyd and M. Warmuth. Sample compression, learnability, and the Vapnik Chervonenkis dimension. *Machine Learning*, 27:1–36, 1995.

Y. Freund. An adaptive version of the boost by majority algorithm. In *Proceedings of the Annual Conference on Computational Learning Theory*, 1999.

Y. Gat. A learning generalization bound with an application to sparse-representation classifiers. *Machine Learning*, 42(3):233–240, 2001.

T. Graepel and R. Herbrich. The kernel Gibbs sampler. In T. K. Leen, T. G. Dietterich, and V. Tresp, editors, *Advances in Neural Information Processing Systems 13*, pages 514–520, Cambridge, MA, 2001. MIT Press.

T. Graepel, R. Herbrich, and K. Obermayer. Bayesian Transduction. In S. A. Solla, T. K. Leen, and K.-R. Müller, editors, *Advances in Neural Information Processing Systems 12*, pages 456–462, Cambridge, MA, 2000a. MIT Press.

T. Graepel, R. Herbrich, and J. Shawe-Taylor. Generalisation error bounds for sparse linear classifiers. In *Proceedings of the Thirteenth Annual Conference on Computational Learning Theory*, pages 298–303, 2000b.

T. Graepel, R. Herbrich, and R. C. Williamson. From margin to sparsity. In T. K. Leen, T. G. Dietterich, and V. Tresp, editors, *Advances in Neural Information Processing Systems 13*, pages 210–216, Cambridge, MA, 2001. MIT Press.

R. Herbrich and T. Graepel. A PAC-Bayesian margin bound for linear classifiers. *IEEE Transactions on Information Theory*, 2002.

R. Herbrich, T. Graepel, and C. Campbell. Bayes point machines. *Journal of Machine Learning Research*, 1:245–279, 2001.

R. Herbrich and R. C. Williamson. Algorithmic luckiness. *Journal of Machine Learning Research*, 3:175–212, 2002.

M. J. Kearns and R. E. Schapire. Efficient distribution-free learning of probabilistic concepts. *Journal of Computer and System Sciences*, 48(3):464–497, 1994.

S. S. Keerthi, S. K. Shevade, C. Bhattacharyya, and K. R. K. Murthy. A fast iterative nearest point algorithm for support vector machine classifier design. Technical Report Technical Report TR-ISL-99-03, Indian Institute of Science, Bangalore, 1999.

G. S. Kimeldorf and G. Wahba. A correspondence between Bayesian estimation on stochastic processes and smoothing by splines. *Annals of Mathematical Statistics*, 41:495–502, 1970.

N. Littlestone and M. Warmuth. Relating data compression and learnability. Technical report, University of California Santa Cruz, 1986.

L. Lovasz. Hit-And-Run mixes fast. *Mathematical Programming A*, 86:443–461, 1999.

D. J. C. MacKay. The evidence framework applied to classification networks. *Neural Computation*, 4(5):720–736, 1992.

D. A. McAllester. Some PAC Bayesian theorems. In *Proceedings of the Annual Conference on Computational Learning Theory*, pages 230–234, Madison, Wisconsin, 1998. ACM Press.

S. Mika, G. Rätsch, J. Weston, B. Schölkopf, and K.-R. Müller. Fisher discriminant analysis with kernels. In Y.-H. Hu, J. Larsen, E. Wilson, and S. Douglas, editors, *Neural Networks for Signal Processing IX*, pages 41–48. IEEE, 1999.

T. M. Mitchell. Generalization as search. *Artificial Intelligence*, 18(2):202–226, 1982.

A. B. J. Novikoff. On convergence proofs on perceptrons. In *Proceedings of the Symposium on the Mathematical Theory of Automata*, volume 12, pages 615–622. Polytechnic Institute of Brooklyn, 1962.

J. Platt. Fast training of support vector machines using sequential minimal optimization. In B. Schölkopf, C. J. C. Burges, and A. J. Smola, editors, *Advances in Kernel Methods—Support Vector Learning*, pages 185–208, Cambridge, MA, 1999. MIT Press.

F. Rosenblatt. The perceptron: A probabilistic model for information storage and organization in the brain. *Psychological Review*, 65(6):386–408, 1958.

B. Schölkopf, R. Herbrich, and A. Smola. A generalized representer theorem. In *Proceedings of the Annual Conference on Computational Learning Theory*, pages 416–426, 2001.

J. Shawe-Taylor, P. L. Bartlett, R. C. Williamson, and M. Anthony. Structural risk minimization over data-dependent hierarchies. *IEEE Transactions on Information Theory*, 44(5):1926–1940, 1998.

R. L. Smith. Efficient Monte-Carlo procedures for generating points uniformly distributed over bounded regions. *Operations Research*, 32:1296–1308, 1984.

A. J. Smola and B. Schölkopf. Sparse greedy matrix approximation for machine learning. In P. Langley, editor, *Proceedings of the International Conference on Machine Learning*, pages 911–918, San Francisco, 2000. Morgan Kaufmann Publishers.

V. Vapnik. *The Nature of Statistical Learning Theory*. Springer, New York, 1995. ISBN 0-387-94559-8.

V. N. Vapnik. *Estimation of Dependences Based on Empirical Data*. Springer, Berlin, 1982.

INDEX

FORTHCOMING VOLUMES

1992–1992: *Control Theory*
   Robotics

1996 Summer Program: *Emerging Applications of Number Theory*

1996–1997: *Mathematics in High Performance Computing*
   Algorithms for Parallel Processing
   Evolutionary Algorithms
   The Mathematics of Information Coding, Extraction and Distribution
   Structured Adaptive Mesh Refinement Grid Methods
   Computational Radiology and Imaging: Therapy and Diagnostics
   Mathematical and Computational Issues in Drug Design
   Rational Drug Design
   Grid Generation and Adaptive Algorithms
   Parallel Solution of Partial Differential Equations

1997 Summer Program: *Statistics in the Health Sciences*
   Week 1: Genetics
   Week 2: Imaging
   Week 3: Diagnosis and Prediction
   Weeks 4 and 5: Design and Analysis of Clinical Trials
   Week 6: Statistics and Epidemiology: Environment and Health

1997–1998: *Emerging Applications for Dynamical Systems*
   Numerical Methods for Bifurcation Problems
   Multiple-time-scale Dynamical Systems
   Dynamics of Algorithms